An Introduction to
Nonlinear Image Processing

Books in the SPIE Tutorial Texts Series

An Introduction to Nonlinear Image Processing

Edward R. Dougherty
Center for Imaging Science
Rochester Institute of Technology

Jaakko Astola
Signal Processing Laboratory
Tampere University of Technology

Donald C. O'Shea, Series Editor
Georgia Institute of Technology

TUTORIAL
TEXTS
IN OPTICAL
ENGINEERING

Volume TT 16

SPIE Optical Engineering Press

A Publication of SPIE—The International Society for Optical Engineering
Bellingham, Washington USA

Library of Congress Cataloging-in-Publication Data

Dougherty, Edward R.
   An introduction to nonlinear image processing / Edward R. Dougherty, Jaakko
Astola.
      p.   cm. — (Tutorial texts in optical engineering ; v. TT 16)
  Includes bibliographical references and index.
   ISBN 0-8194-1560-X
   1. Image processing — Digital techniques — Mathematics.   2. Filters
(Mathematics)   3. Nonlinear theories.   I. Astola, Jaakko.
II. Title.   III. Series.
TA1637.D68   1994
621.36'7'0151—dc20   93-48070
                                            CIP

Published by

SPIE—The International Society for Optical Engineering
P.O. Box 10
Bellingham, Washington 98227-0010

Printed in the United States of America

*Cover illustration: (top left) Ideal circle image. (top right) Noisy circle image.
(bottom left) Circle image restored by classical pruners. (bottom right) Circle
image restored by optimal binary filters.*

Introduction to the Series

These Tutorial Texts provide an introduction to specific optical technologies for both professionals and students. Based on selected SPIE short courses, they are intended to be accessible to readers with a basic physics or engineering background. Each text presents the fundamental theory to build a basic understanding as well as the information necessary to give the reader practical working knowledge. The included references form an essential part of each text for the reader requiring a more in-depth study.

Many of the books in the series will be aimed at readers looking for a concise tutorial introduction to new technical fields, such as CCDs, sensor fusion, computer vision, or neural networks, where there may be only limited introductory material. Still others will present topics in classical optics tailored to the interests of a specific audience such as mechanical or electrical engineers. In this respect the Tutorial Text serves the function of a textbook. With its focus on a specialized or advanced topic, the Tutorial Text may also serve as a monograph, although with a marked emphasis on fundamentals.

As the series develops, a broad spectrum of technical fields will be represented. One advantage of this series and a major factor in the planning of future titles is our ability to cover new fields as they are developing, giving people the basic knowledge necessary to understand and apply new technologies.

Donald C. O'Shea   January 1994
Georgia Institute of Technology

To our wives
Terry and Ulla

Table of Contents

Preface

From a strict semantic point of view, nonlinear image processing encompasses all image processing that is not based on linear operators; however, from a practical, evolutionary point of view, the name itself is usually associated with the study of nonlinear filters, mainly the deterministic and nondeterministic analysis and design of logic-based operators. We shall take the latter perspective, the emphasis being on representation, design, and statistical optimization of nonlinear filters. Even with this restriction, the subject matter is extensive and growing more rapidly than ever. We shall, for the most part, remain at a fundamental level, covering basic filters.

Historically, nonlinear image processing has grown from within three, interdependent orientations. Filters have been based on geometric structure (morphological operators), numerical ordering (median-type filters), and logic (binary and stack filters). As will be seen in the text, the geometric and logical approaches are primary and, for windowed filters, equivalent. We could, therefore, take the mathematical approach that all filtering is rooted in geometry or logic. While unification has mathematical benefits, each of the three strands has its own advantages relative to applications, especially in the different insights upon which each draws. Thus, while we discuss the unifying representations, we maintain the historical evolution and treat all three approaches in their own rights.

After an introduction, the book begins with the basic operations of binary morphological image processing and then turns to application of morphological openings and closings. The fourth chapter treats unification of structural morphological operations and discusses their logical roots. Gray-scale morphological operations are discussed next. The book then turns to medians, order statistics, and stack filters. Emphasis is placed on both the statistical roots in maximum-likelihood estimation and the logical roots in threshold decomposition. Optimization of median-type filters is also discussed. The tenth chapter treats statistical optimization of increasing and nonincreasing binary filters. For those interested mainly in this topic, one could go directly from the fourth to the tenth chapter. Finally, in the last chapter we provide brief descriptions of some nonlinear filters that we believe rest upon sound mathematical foundations.

We acknowledge and offer our appreciation to all who assisted in preparation of the book. These include I. Korkee, who organized and prepared the manuscript, and Y. Chen, C. Cuciurean-Zapan, J. Pelz, A. Wiseman, R. Loce, H. Huttunen, S. Siren, and R. Yang, who contributed figures and images. We offer our appreciation to H. Longbotham, who technically reviewed the manuscript, and to E. Pepper, who provided editorial assistance and who guided the manuscript through production.

Edward R. Dougherty   January 1994
Jaakko Astola

Chapter 1

Introduction

Since digital image processing involves the processing of finite bit strings through logic circuits, in principle every algorithm possesses a logical representation over a finite number of binary variables. Two conclusions are immediate: (1) from both the algebraic and operational perspectives, representations of logical operators lie at the base of digital image processing, and (2) all image processing is nonlinear. The consequences of the first conclusion will be evident throughout the text, namely, the subject matter remains within a logical-variable context. The meaning of the second conclusion is a bit more subtle. If all image processing is nonlinear, then why differentiate nonlinear image processing from image processing in general? The answer has two aspects: historical and practical.

Historically, many two-dimensional methods of image processing have grown out of the one-dimensional methods of signal processing. The latter are, to a great extent, linear. Since linear methods are well understood, easy in comparison to nonlinear methods, fairly tractable, and, in many circumstances, proven to be effective, it is only natural that early researchers should have turned to them. In hindsight, it is easy to see why various problems have not succumbed to linear methods; however, it is also clear that linear methods have also proven to be very successful at solving certain classes of imaging problems. Since the popularity of specifically nonlinear approaches has been more recent, it is pedagogically sound to differentiate between linear and nonlinear methods.

Many common phenomena (sound, ac voltage, seismic waves, etc.) are by nature exponentials and are therefore eigenfunctions of linear systems, thereby making linear filtering quite natural. Moreover, in many applications, such as communications, noise can be modeled as Gaussian and again linear filtering is appropriate. Consequently, for many signal processing applications there is no need to consider nonlinear methods and it has been among these that most modern digital signal processing has been developed.

With images, matters are completely different. Images are often hard to model as wide-sense stationary processes, they do not reside on a narrow subband of the usable frequency range, and the frequency characterization of the noise or corrupting process may be very similar to that of the image itself.

Many classical problems, even when very advanced signal processing is used, resolve to simple estimation or detection tasks: is a specific bit one or zero, or what is the frequency of the signal? In image processing we are often interested in more complex tasks, such as discovering size distributions of objects or classifying visible objects into different classes. If the task is to enhance an image for human viewing, it is not clear that noise suppression is the primary goal. This intuitively means that we cannot project the problem into some suitable small-dimensional space, a common feature of many linear methods.

Linear methods often have significant practical advantages. These tend to arise from orthogonality within linear spaces. The various Fourier-type representations rely on the existence of orthogonal bases and these lead to both simple computational expressions and simple invertible transforms. In particular, optimal mean-square-error linear filters (for enhancement, restoration, compression, etc.) are straightforward to derive: for a finite number of observations one is confronted only with a finite system of linear equations and for an infinite number of observations, under the assumption of wide-sense stationarity, the problem is transformed to an easy one in the frequency domain. Such straightforward, elegant approaches are rarely possible when working with nonlinear filters.

All image processing is, of necessity, nonlinear. Why then have linear methods proven fruitful when the mathematics behind these methods implicitly assume the objects under consideration lie in a vector space, and such spaces cannot be represented in finite logic circuits? The answer is straightforward: various theorems and algorithms exist to both describe the effects of digitization and provide approximate digital formulations of continuous (nondigital) expressions.

Yet, given the natural relationship between digital computation and logical representation, might it not be advantageous to ground digital image processing directly in logic: not in the sense that one wishes to throw out linear methods, but rather to see if a direct approach leads to useful inherent nonlinear operations? Practically, the restriction to linearity imposes a severe constraint on the type of algorithms one can employ. If the cost of easy mathematical tractability (the key advantage of linearity) is the inability to successfully accomplish key image processing tasks or to accomplish them with undesirable side effects, then the cost is too high and one must be prepared to confront more difficult mathematical problems. A key example in this regard is noise reduction via optimal filters. Design of an optimal smoothing linear filter is straightforward; however, often there is an unacceptable blurring of important image information, so much so that a simple (nonoptimized) median filter can provide superior results. The conclusion is evident: although the problem may be difficult, it is incumbent that we try to design statistically optimal nonlinear filters. Owing to the nature of digital processing, optimization will be over classes of logical operators, although for the sake of conceptualization and mathematical formulation an optimization problem may be posed in terms of shape or numerical ordering.

As a subject, nonlinear image processing has tended to focus on the design and analysis of filters, often with the idea that these filters will be used for restoration (suppression of noise). More specifically, we are interested in classes of filters, such as filters that are spatially invariant, increasing, or idempotent, these being examples of algebraic properties that an operator may or may not satisfy. From a statistical perspective, a filter is an estimator: it operates on a set (window) of observations and produces an estimate of an unobserved quantity. A linear filter is a linear combination of the observations; a nonlinear filter (in digital image processing) is a logical combination of the observations. Although nonlinear techniques abound in all areas of image processing, we shall remain within the traditional confines of the subject (as it has come to be known) and focus on the design, analysis, and application of nonlinear filters.

From the standpoint of applications, the salient advantage of nonlinear filters is their ability to pass structural information while suppressing noise or removing clutter. Pattern and edge information are often crucial to image understanding, and in many circumstances it is possible to design nonlinear filters that pass structural information in a manner superior to that of linear filters. Just as the weights of a linear filter are tuned to a specific image environment, design of nonlinear filters and the application of certain classes of nonlinear filters are often application-specific.

Nonlinear filtering has developed along three lines: logical, geometrical, and numerical. These can be reformulated as set-, shape-, and order-based. While these three approaches have often proceeded in parallel without much interaction, it will become clear from the text that they are deeply interrelated. Owing to digital circuitry, logic is at the foundation, but owing to phenomenology, algorithms and analysis have developed around shape and numerical ordering.

Shape-based nonlinear filtering is centered around mathematical morphology, which, according to its formulation by Matheron [123], is grounded in the Minkowski operations as incorporated by Hadwiger [71]. Here there are very primitive set operations, erosion and dilation, which, in conjunction with the basic set operations, form the building blocks of higher-level operations, which themselves have historically been used in a shape-based mode. Nonetheless, while the primitive erosion and dilation operations have been viewed from a pattern perspective, which is quite natural since they have been used for processing images, they are, at their foundation, logically based.

This was clearly recognized at the outset by Matheron. He observed that the binary morphological operations with which he was concerned (erosion, dilation, opening, closing, their iterations, and their combination under unions and intersections) are increasing, translation-invariant set operations. When implemented digitally, they become increasing, windowed operators and therefore possess logic-gate sum-of-product (maximum-of-minima) representations in terms of noncomplemented logical variables. Erosion is the key operation of nonlinear image processing because in a finite-window setting, erosion is simply a product of uncomplemented logical variables. Recognizing this, Matheron set down the algebraic epistemology of nonlinear image processing in his celebrated representation theorem: every increasing, translation-invariant set operator possesses a representation as a union of erosions. The proof of the theorem is not difficult and there have been many extensions; nevertheless, the key insight is that the shape-based morphological operations are grounded in a theory of sets that reduces to classical logic representation in a finite digital setting.

The other strand of nonlinear image processing has developed around order-based filters, the prototype being the median. Order-based statistics have proven very useful in distribution-free statistical decision theory. In image processing, they have been employed successfully to pass desired image structure while at the same time suppressing noise, and, for the most part, they have been used on gray-scale images. Operationally, order-based filters are expressed in terms of logical variables; that is, every order-based (gray-scale) filter possesses a sum-of-products expression in terms of binary logical variables. Practically, binary representation can be achieved by stacking gray-scale slices of the image and operating in a binary fashion on each slice, thereby getting at once to the Matheron representation in the case of increasing operators (including order-statistic filters). As must be the case, logic is at the base; however, in analogy to morphological processing, there is great benefit from viewing order-based filters numerically, rather than at their lowest processing level.

Gazing ahead to all the images in the book, a practical person might wonder why we have begun by discussing mathematical and logical questions surrounding nonlinear image processing. Paradoxically, we have done so to smooth the linear-processor's transition to nonlinear processing. Linearity has proved useful for many applications; nevertheless, it represents a mathematical constraint that presupposes the kind of processing one can perform. Nonlinearity represents a removal of that constraint. This does not mean that we proceed without constraints; rather, it means we impose different constraints, ones that are either more appropriate to a particular application or ones that are less constraining overall. In the first instance, if shape information is key, we apply shape-based constraints; in the latter, we achieve greater latitude and a wider universe of possible solutions. Nonetheless, there is the ever-present key constraint of finite-logical-variable processing. This constraint poses subtle problems for linear processing; nonlinear processing takes this logical environment as its natural mathematical framework. If one keeps this in mind, then the methods of nonlinear image processing appear natural and unified, and their success can be more readily appreciated.

Chapter 2

Basic Binary Operations

From a mathematical perspective, binary nonlinear image processing is one and the same with binary mathematical morphology; however, since our intention is to focus on operators and their properties, we will consider the binary morphological operational theory to constitute the subject matter of binary nonlinear image processing, not the various special-purpose algorithms that have historically composed the subject matter of binary morphological image processing (watersheds, skeletons, etc.). This chapter introduces the basic operators of binary mathematical morphology, with emphasis on their definitions and geometric interpretations.

2.1. Introduction

From a geometric perspective, the most basic morphological idea is to probe an image with a **structuring element** and to mark locations at which the structuring element fits within the image. By marking the locations at which a structuring element fits within the image, we derive structural information concerning the image. This information depends on both the size and shape of the structuring element, and, as emphasized by Matheron, the nature of that information is therefore dependent on the choice of the structuring element. To paraphrase Matheron, as is typically the case with scientific knowledge the knowledge we have concerning an image is relative to the manner in which we probe (observe) it, and all consequent relationships are dependent on our choices regarding the manner of our observations. Even if we apply machine algorithms to select appropriate structuring elements, the criteria by which the algorithms make their selections are ultimately determined by the kind of information we desire.

We will consider two types of binary images, Euclidean and discrete. A Euclidean binary image is a subset of n-dimensional Euclidean space. For signal processing, $n = 1$, and for image processing, $n = 2$. For the most part we will focus on image processing, so a Euclidean image will be a subset of the Euclidean plane. For digital implementation we will consider an image to be a subset of the two-dimensional Cartesian grid. It is important to recognize the twofold nature of nonlinear image processing: it is fundamentally both geometric and logical in character. Geometric appreciation is rooted in Euclidean space; logical appreciation (and implementation) is rooted in discrete space.

2.2. Erosion

Characterization of fitting depends on two concepts, subsethood and Euclidean-space translation. The **translation** of a set A by a point x is denoted by $A + x$ and is defined by

$$A + x = \{a + x : a \in A\}. \tag{2.1}$$

Geometrically, $A + x$ is A translated along the vector x. The nature of probing is to mark the positions (translations) of a structuring element where it fits into an image.

The morphological erosion operation serves as the mathematical marker of structuring element fits within an image. The **erosion** of set A by set (structuring element) B is denoted by $A \ominus B$ and is defined by

$$A \ominus B = \{x : B + x \subset A\}. \tag{2.2}$$

Erosion is also denoted by $E(A, B)$. $A \ominus B$ consists of all points x for which the translation of B by x fits inside of A. Treating B as a template, $A \ominus B$ consists of all template origin positions for which the translated template fits inside A.

Figure 2.1 shows a circular structuring element translated to two different locations in the image A. Since $B + x \subset A$, $x \in A \ominus B$; since $B + z \not\subset A$, $z \notin A \ominus B$.

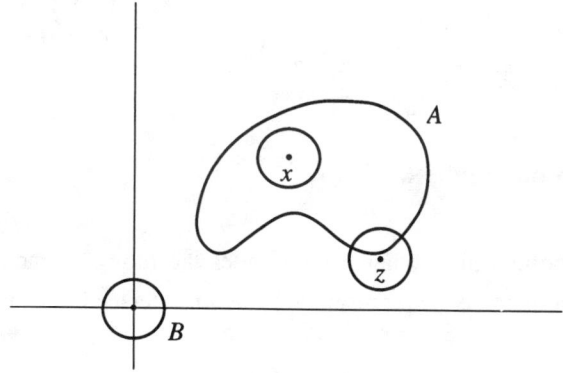

Fig. 2.1. A structuring element fitting and not fitting.

If the origin lies inside of the structuring element, then erosion has the effect of shrinking the input image. This is illustrated in Fig. 2.2, where the structuring element B is a disk. Geometrically, the disk B has been moved around inside of A and the positions of the origin (in this case the center of the disk) have been marked so as to produce the eroded image. Formally, we can state the following property: if the origin is contained within the structuring element, then the eroded image is a subset of the input image. Should the origin not lie within the structuring element, then it may not be that the eroded image lies within the input image. This situation is illustrated in Fig. 2.3.

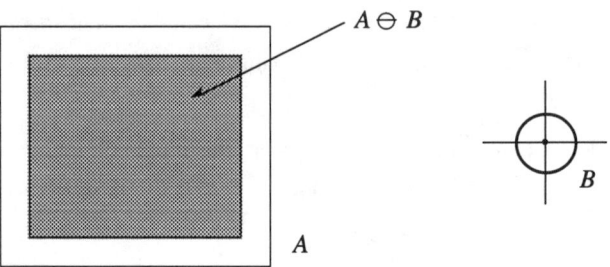

Fig. 2.2. Erosion as shrinking.

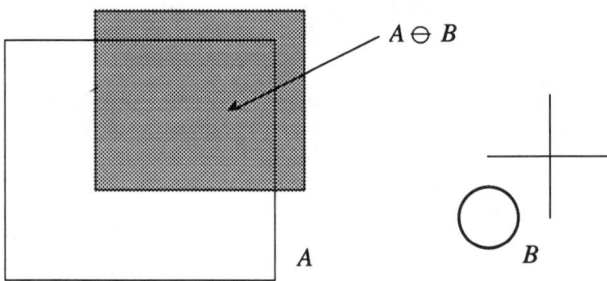

Fig. 2.3. Erosion not a subset.

Erosion can be formulated in other ways besides the fitting characterization of Eq. 2.2. Of particular importance is its representation by an intersection of image translates:

$$A \ominus B = \bigcap \{A - b : b \in B\}. \tag{2.3}$$

Here, the erosion is found by intersecting all translates of the input image by negatives of points in the structuring element. The method is illustrated in Fig. 2.4. While the fitting definition of erosion is paramount for image-processing insight, the formulation of Eq. 2.3 is theoretically useful.

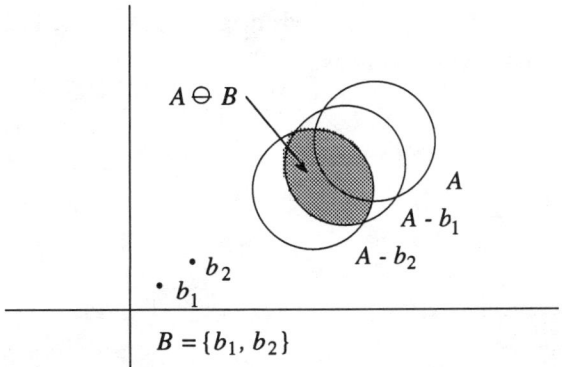

Fig. 2.4. Erosion as an intersection of translates.

The erosion formulation of Eq. 2.3 is closely related to a classical set operation first studied by Minkowski: relative to erosion and its formulation via Eq. 2.3, **Minkowski subtraction** of A by B is defined by

$$A \ominus (-B) = \bigcap \{A + b : b \in B\}, \tag{2.4}$$

where $-B = \{-b : b \in B\}$ is the reflection of B through the origin or, equivalently, the rotation of B about the origin. In words, Minkowski subtraction is erosion by the rotated structuring element. When reading the literature, one must show care with regard to the notation $\ominus$, since in many cases it refers to Minkowski subtraction and erosion is defined relative to Minkowski subtraction. Here we stay with current practice and use $\ominus$ to denote erosion.

The fitting characterization of Eq. 2.2 applies directly to digital space (as does the intersection formulation of Eq. 2.3). Consider the digital image

$$S = \begin{pmatrix} 0 & 1 & 0 & 1 & 0 \\ 0 & 1 & 1 & 0 & 1 \\ \mathbf{0} & 1 & 1 & 1 & 0 \end{pmatrix}, \tag{2.5}$$

where the bold value is at the grid origin and the remaining values have grid locations relative to the origin value. Consider also the structuring element

$$E = \begin{pmatrix} 1 & 0 \\ 1 & 1 \end{pmatrix}. \tag{2.6}$$

As illustrated in Fig. 2.5, where the input image is first depicted alone and then with a translated copy of the structuring element within it, translating E by $(2, 1)$ yields a fit, thereby showing that $(2, 1)$ lies in the eroded image. Marking all fits of the translated structuring element yields the eroded image

$$S \ominus E = \begin{pmatrix} 0 & 0 & 1 & 0 \\ \mathbf{0} & 0 & 1 & 1 \end{pmatrix}. \tag{2.7}$$

S Translated
                                 Structuring
                                 Element

Fig. 2.5. Digital erosion.

We have mentioned that when the structuring element contains the origin, the eroded image is a subset of the input image. An important case to the contrary is when erosion is used to fill holes in an image. The image

$$S = \begin{pmatrix} 1 & 1 & 1 & 1 & 1 & 1 & 1 \\ 1 & 1 & 0 & 1 & 1 & 1 & 1 \\ 1 & 1 & 1 & 1 & 0 & 1 & 1 \\ 1 & 0 & 1 & 0 & 1 & 1 & 1 \\ \mathbf{1} & 1 & 1 & 1 & 1 & 0 & 1 \end{pmatrix} \tag{2.8}$$

might represent a 5 by 7 square with some salt holes. Eroding S by $E = \begin{pmatrix} 1 & \mathbf{0} & 1 \end{pmatrix}$ yields

$$S \ominus E = \begin{pmatrix} 0 & 1 & 1 & 1 & 1 & 1 & 0 \\ 0 & 0 & 1 & 0 & 1 & 1 & 0 \\ 0 & 1 & 1 & 0 & 1 & 0 & 0 \\ 0 & 1 & 0 & 1 & 0 & 1 & 0 \\ \mathbf{0} & 1 & 1 & 1 & 0 & 1 & 0 \end{pmatrix}. \tag{2.9}$$

Restoration of the full 5 by 7 square is given by $S \cup (S \ominus E)$.

An appealing feature of mathematical morphology is the existence of a system of algebraic relations involving its basic operations and the basic set-theoretic operations. Taken together, these relations constitute the **Minkowski algebra**. Here we will be mentioning only the most fundamental algebraic properties, those having a direct impact on the operational theory of nonlinear image processing, and defer to a text on morphological image processing for a more complete account.

A key property of erosion is that, as an operation, it commutes with intersection:

$$\left(\bigcap_i A_i \right) \ominus B = \bigcap_i A_i \ominus B. \tag{2.10}$$

Intersecting a collection of images and then eroding is equivalent to eroding each and then intersecting. Eq. 2.10 plays a central role in the theoretical development of nonlinear image processing within the context of lattice theory. Its proof is immediate from the definition of erosion: x lies in the intersection of all the erosions $A_i \ominus B$ if and only if $B + x$ is a subset of every A_i, but this is equivalent to $B + x$ being a subset of the intersection of the A_i.

2.3. Dilation

Dual to erosion, meaning it is defined via erosion by complementation, is **dilation** of set A by set B. It is denoted by $A \oplus B$ and defined by

$$A \oplus B = [A^c \ominus (-B)]^c, \tag{2.11}$$

where A^c denotes the set-theoretic complement of A. Dilation is also denoted by $D(A, B)$. As indicated by Fig. 2.6, where B is a disk, if B contains the origin, then dilation of A by B results in an expansion of A. Since dilation involves a fitting into the complement of an image, it represents a filtering on the outside, whereas erosion represents a filtering on the inside.

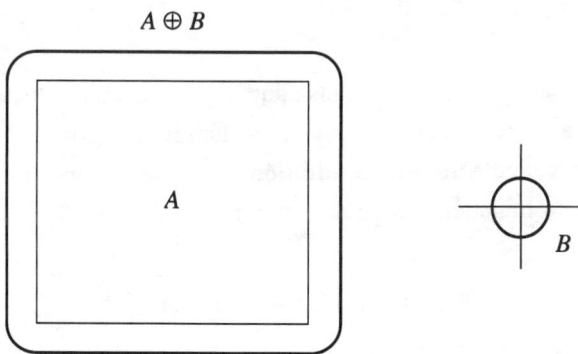

Fig. 2.6. Dilation as expansion.

If we apply Eq. 2.11 to the dilation of A^c by $-B$ (instead of A by B) and then take the complement of each side [recognizing that $(A^c)^c = A$ and $-(-B) = B$], we obtain

$$A \ominus B = [A^c \oplus (-B)]^c, \tag{2.12}$$

which says that erosion can be expressed in terms of dilation (as dilation is expressed in terms of erosion).

Relative to processing images, it is customary to write the input image first and the structuring element second (although there is no mathematical requirement to do so). If we think of a disk structuring element, then dilation fills in small (relative to the disk) holes and protrusions into the image, whereas erosion eliminates small components and extrusions of the image into its complement.

Dilation is both commutative,

$$A \oplus B = B \oplus A, \tag{2.13}$$

and associative,

$$A \oplus (B \oplus C) = (A \oplus B) \oplus C. \tag{2.14}$$

Another formulation of dilation deserves mention. First,

$$A \oplus B = \bigcup \{A + b : b \in B\} \tag{2.15}$$

so that the dilation can be found by translating the input image by all points in the structuring element and then taking the union. Written in the form of Eq. 2.15, dilation has historically been called **Minkowski addition**. Because dilation is commutative, the Minkowski-addition formulation can be rewritten in a geometrically intuitive form:

$$A \oplus B = \bigcup \{B + a : a \in A\}. \tag{2.16}$$

As illustrated in Fig. 2.7, the structuring element is translated to all points in the input image and the union of these translates is taken.

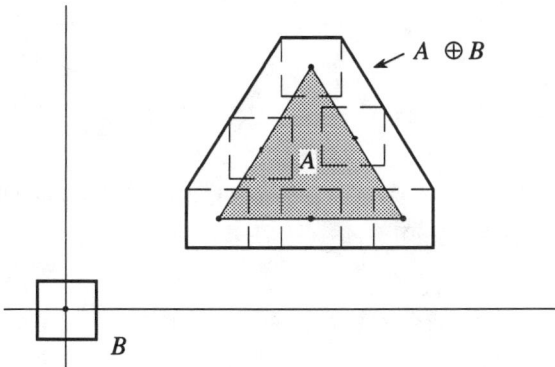

Fig. 2.7. Minkowski addition.

Using the image of Eq. 2.5 and the structuring element of Eq. 2.6, we illustrate digital
dilation using the Minkowski-addition formulation of Eq. 2.16. The structuring element
is translated to all image pixels and the resulting structuring-element translations are
unioned. For instance, since $(1, 2)$ is an activated pixel of S, the structuring element
is translated by $(1, 2)$, the effect being to place the structuring-element origin at $(1, 2)$.
Owing to this translation, pixels $(1, 2)$, $(0, 2)$, and $(0, 3)$ are activated in the dilated image.
Upon forming the union of all such translations, the dilated image is

$$S \oplus E = \begin{pmatrix} 1 & 0 & 1 & 0 & 0 \\ 1 & 1 & 1 & 1 & 0 \\ 1 & 1 & 1 & 1 & 1 \\ 1 & 1 & 1 & 1 & 0 \end{pmatrix}. \tag{2.17}$$

Dual to the commutativity of erosion with intersection, Eq. 2.10, is the commutativity
of dilation with union:

$$\left(\bigcup_i A_i \right) \oplus B = \bigcup_i A_i \oplus B. \tag{2.18}$$

Unioning a collection of images and then dilating is equivalent to dilating every image
and then forming the union.

2.4. Opening and Closing

Erosion and dilation are the primitive building blocks of binary morphological (nonlinear)
processing; we now consider secondary operations that are central to many applications.
The **opening** of image A by image B is denoted by $A \circ B$ and is defined as an iteration

of erosion and dilation by

$$A \circ B = (A \ominus B) \oplus B. \tag{2.19}$$

Other notations for opening are $O(A, B)$, $\gamma_B(A)$, and A_B. A more geometrically intuitive formulation of opening is given by

$$A \circ B = \bigcup \{B + x : B + x \subset A\}. \tag{2.20}$$

Here, the opening results from unioning all translations of the structuring element that fit inside the input image. Each fit is marked and the opening results from taking the union of the structuring-element translations to each marked location. This is precisely what is meant by eroding and then dilating.

Whereas the position of the structuring element affects erosion and dilation, from Eq. 2.20 it can be seen that structuring-element position does not affect the opening: if B_1 and B_2 are translationally congruent, that is there is an x such that $B_1 = B_2 + x$, then $A \circ B_1 = A \circ B_2$.

The expression of opening as erosion followed by dilation is illustrated in Fig. 2.8, where a rectangle is successively eroded and dilated by a disk. It is also possible to discern the effect of fitting, as expressed in Eq. 2.20: opening the rectangle has resulted in it being rounded from the inside, this rounding resulting from the manner in which the disk has been "rolled around" inside the rectangle to achieve a union of the fits. Had the structuring element been a small square with horizontal base, there would have been no rounding and the opened image would have been the same as the original.

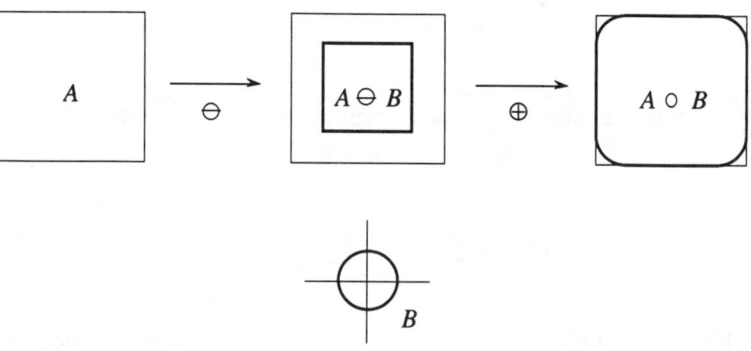

Fig. 2.8. Opening.

Two applications of opening are evident in Fig. 2.8. Opening by a disk results in a filter that smooths from the inside: it rounds corners extending into the background. The roundness of the disk yields a "lowpass" effect; the effect is quite different with a square structuring element.

Rather than view the opened image itself as the final output of the processing, we can take a different view. We can consider the set-theoretic subtraction $A - (A \circ B)$. In Fig. 2.8, this image consists of input-image corners that protrude into the background, and can be employed for recognition purposes. Use of a disk is common because its shape effect is rotationally invariant; however, there are many instances when it is beneficial to employ other kinds of structuring elements.

As for a digitital example, if S and E are the image and structuring element of Eqs. 2.5 and 2.6, respectively, then

$$S \circ E = \begin{pmatrix} 0 & 1 & 0 & 0 \\ 0 & 1 & 1 & 0 \\ \mathbf{0} & 1 & 1 & 1 \end{pmatrix}. \tag{2.21}$$

If we view S as a triangle that has some background noise, then opening by E has restored the triangle. We will shortly have much more to say on this type of restoration.

The dual operation to opening is closing, which is defined as an erosion followed by a dilation. The **closing** of A by B is denoted by $A \bullet B$ and is defined by

$$A \bullet B = [A \oplus (-B)] \ominus (-B). \tag{2.22}$$

Closing is also denoted by $C(A, B)$, $\phi_B(A)$, and A^B. It is dual to opening:

$$A \bullet B = (A^c \circ B)^c. \tag{2.23}$$

Like opening, closing is not affected by structuring-element position. Replacing A by A^c in Eq. 2.23 and complementing yields

$$A \circ B = (A^c \bullet B)^c. \tag{2.24}$$

Figure 2.9 illustrates closing. In it, since B is a disk, rotation plays no role. Rather than employ the iteration of Eq. 2.22, we could employ duality in conjunction with the union formulation of opening given in Eq. 2.20, thereby fitting, or "rolling the ball," around the outside of the image. The effect can be seen in the manner in which the closing has filtered from the outside, smoothing only corners that protrude into the image.

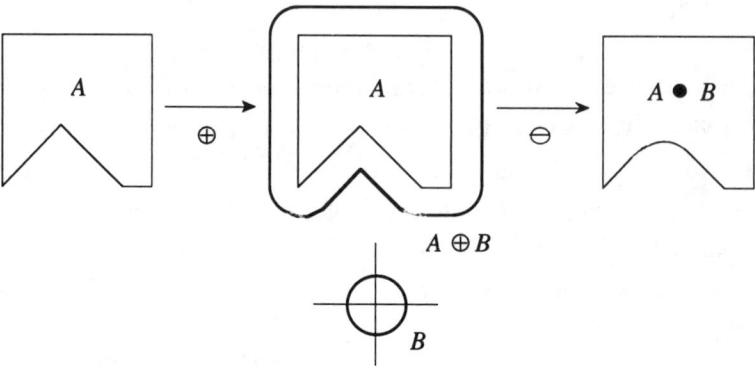

Fig. 2.9. Closing.

For a digital example we close image S of Eq. 2.5 by structuring element E of Eq. 2.6 to obtain

$$S \bullet E = \begin{pmatrix} 0 & 1 & 1 & 1 & 0 \\ 0 & 1 & 1 & 1 & 1 \\ \mathbf{0} & 1 & 1 & 1 & 0 \end{pmatrix}. \qquad (2.25)$$

Chapter 3

Binary Filtering with Opening and Closing

A great number of image processing tasks can be accomplished using openings and closings. Among these are many tasks typically associated with filtering, in particular, restoration. The present chapter examines opening and closing from the perspective of certain key nonlinear filter properties and discusses various filtering applications.

3.1. Filter Properties

Certain fundamental notions concerning image operators (filters) are useful for the description of basic properties pertaining to the operators of nonlinear image processing and they will be key to our discussions regarding nonlinear filtering. At present our concern is with operations on binary images that yield binary images, as opposed to those that yield gray-level images or those that yield numerical values. If A is the input image, we let $\Psi(A)$ denote the output image. For instance, if Ψ is erosion by B, then $\Psi(A) = A \ominus B$; if Ψ is dilation by B, then $\Psi(A) = A \oplus B$. Throughout the book we will be considering many operators and Ψ will therefore take many forms. To fully appreciate the algebraic structure of nonlinear image processing, especially as it relates to filtering, one must recognize that there are certain properties that an operator may or may not possess that make it useful or not useful for a certain task. For instance, in linear processing operators are required to be linear, and the degree to which linearity is or is not an appropriate constraint determines the efficacy of it being required. Here we consider operator properties relevant to nonlinear processing.

An operator Ψ is said to be **translation invariant** if translating the input image and then operating by Ψ is equivalent to operating by Ψ and then translating:

$$\Psi(A + x) = \Psi(A) + x. \tag{3.1}$$

Dilation, erosion, opening, and closing are translation invariant:

$$(A + x) \oplus B = (A \oplus B) + x \tag{3.2}$$

$$(A + x) \ominus B = (A \ominus B) + x \tag{3.3}$$

$$(A + x) \circ B = (A \circ B) + x \tag{3.4}$$

$$(A + x) \bullet B = (A \bullet B) + x. \tag{3.5}$$

For dilation, this means that the same output results from first translating the image and then dilating by a given structuring element as would result by first dilating the image by the structuring element and then translating. Analogous comments apply to erosion, opening, and closing.

When considering translation invariance, one must be careful to recognize that it applies to translating the image, not the structuring element. Nevertheless, because dilation is commutative, it follows from Eq. 3.2 that

$$A \oplus (B + x) = (A \oplus B) + x \tag{3.6}$$

so there is translation invariance relative to the structuring element. More care must be taken with erosion; in fact,

$$A \ominus (B + x) = (A \ominus B) - x \tag{3.7}$$

so that translating the structuring element prior to eroding is equivalent to eroding and then translating the eroded image by the same amount in the opposite direction. For both opening and closing, only the shape of the structuring element affects the operation, not the position. Thus,

$$A \circ (B + x) = A \circ B \tag{3.8}$$

$$A \bullet (B + x) = A \bullet B. \tag{3.9}$$

An operator Ψ is said to be **monotonically increasing** if A_1 a subset of A_2 implies $\Psi(A_1)$ a subset of $\Psi(A_2)$, so that Ψ preserves order. For a fixed structuring element, dilation, erosion, opening, and closing are increasing: if $A_1 \subset A_2$, then $A_1 \oplus B \subset A_2 \oplus B$, $A_1 \ominus B \subset A_2 \ominus B$, $A_1 \circ B \subset A_2 \circ B$, and $A_1 \bullet B \subset A_2 \bullet B$.

Increasing monotonicity for erosion is relative to a fixed structuring element and input images ordered by set inclusion; quite a different phenomenon occurs if the input image is kept constant and two ordered structuring elements are employed. If A is a fixed image and B_1 is a subset of B_2, then B_1 fits inside of A whenever B_2 fits. Consequently, $A \ominus B_1$ contains $A \ominus B_2$. This property is central to the design of filters involving multiple erosions.

We have mentioned that dilation is dual to erosion and closing is dual to opening. The notion of duality is central to nonlinear processing. In general, given an operator Ψ, the **dual operator** is denoted by Ψ^* and is defined by

$$\Psi^*(A) = \Psi(A^c)^c. \tag{3.10}$$

According to Eqs. 2.11 and 2.23, dilation by B is the dual of erosion by $-B$ and closing by B is the dual of opening by B.

In general, the dual of the dual is the original operator, $\Psi^{**} = \Psi$. This relation follows directly from the definition:

$$\Psi^{**}(A) = \Psi^*(A^c)^c = \Psi(A^{cc})^{cc} = \Psi(A). \tag{3.11}$$

Because $\Psi^{**} = \Psi$, duality expressions occur in pairs, Ψ^* as the dual of Ψ and Ψ as the dual of Ψ^*. The duality pair for erosion and dilation is composed of Eqs. 2.11 and 2.12; the duality pair for opening and closing is composed of Eqs. 2.23 and 2.24.

An operator Ψ is said to be **antiextensive** if $\Psi(A)$ is always a subset of A, and it is said to be **extensive** if $\Psi(A)$ always contains A. Opening is antiextensive: $A \circ B \subset A$. This follows from the fact that the opening is a union of translates lying within the input image. Closing is extensive: $A \bullet B \supset A$.

An operator Ψ is said to be **idempotent** if, for any set A, $\Psi(\Psi(A)) = \Psi(A)$. In words, operating twice by Ψ is equivalent to operating once by Ψ. Both opening and closing are idempotent:

$$(A \circ B) \circ B = A \circ B \tag{3.12}$$

$$(A \bullet B) \bullet B = A \bullet B. \tag{3.13}$$

The import of idempotence is that once an image has been opened (closed), subsequent openings (closings) produce no further effects. This is quite different than erosion or, if we think of linear processing, moving-average filters.

In sum, for the elementary morphological operations, erosion and dilation are translation invariant and increasing; opening (closing) is also antiextensive (extensive) and idempotent.

3.2. Application of Opening and Closing Filters

We examine the kind of restoration that can be effected by opening and closing. Consider the text image of Fig. 3.1 and the degraded version of it shown in Fig. 3.2, the degradation arising from pepper (union) noise. Opening by a 3×3 (pixel) square produces the filtered image of Fig. 3.3. Opening by the square has a restorative effect because the square does not fit into most of the small noise components strewn about the background of the text image, thereby eliminating those into which it does not fit. Restoration is not perfect: some detail of the original characters is lost because the structuring element does not fit in such a way as to pass every pixel in the original characters and some noise remains because noise pixels are part of components into which the structuring element fits.

Owing to open-close duality, analogous remarks apply to closing filtering. To illustrate, we again consider the text image of Fig. 3.1; however, this time we degrade it by salt (subtractive) noise. The degraded image is shown in Fig. 3.4 and the closing with a 2×2 structuring element is shown in Fig. 3.5.

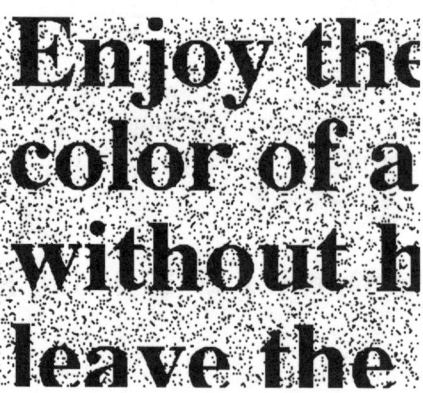

Fig. 3.1. Ideal text image.

Fig. 3.2. Text image degraded by pepper noise.

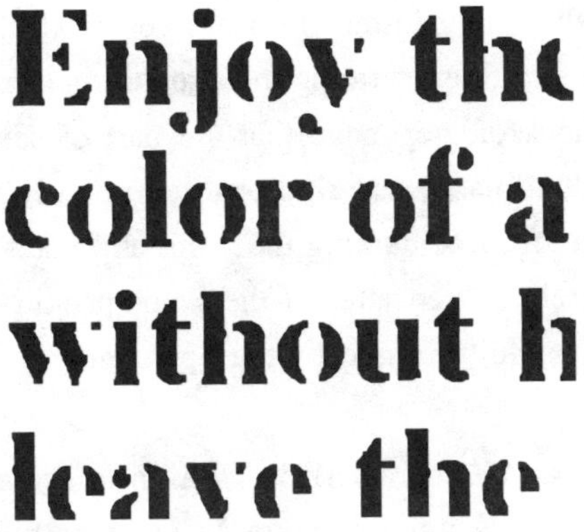

Enjoy the color of a without h leave the

Fig. 3.3. Opened pepper-degraded text image.

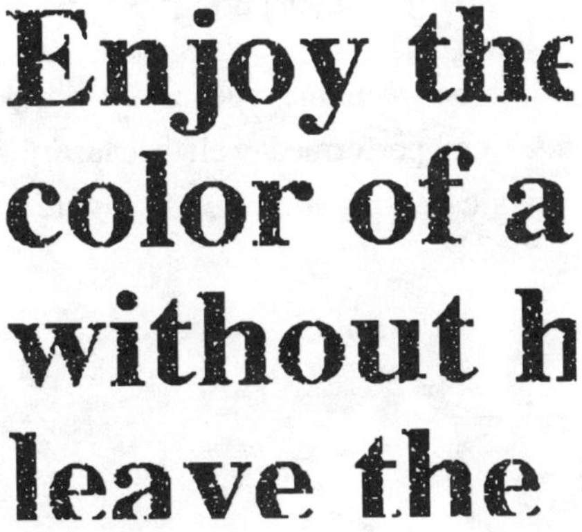

Enjoy the color of a without h leave the

Fig. 3.4. Text image degraded by salt noise.

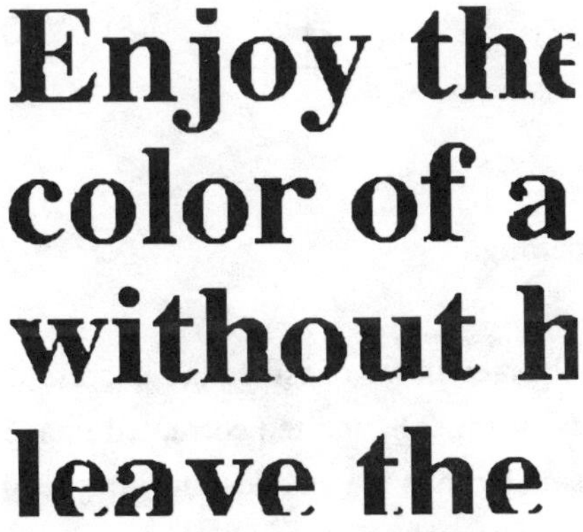

Enjoy the color of a without h leave the

Fig. 3.5. Closed salt-degraded text image.

Careful consideration of the fitting formulation of opening (Eq. 2.20) shows the manner in which the opening acts as a filter: treating the structuring element as a shape primitive, it passes only those portions of the image that are part of some translate of the shape primitive that fits inside the image. Put rigorously, a point lies in $A \circ B$ if and only if there exists some translate of B containing the point and itself being contained in A. If an image is made up entirely of translates of the shape primitive (that is, if it is a union of such translates), then it is fully passed by the opening.

The filtering of Fig. 3.3 can be characterized with an image-noise model. There is an underlying uncorrupted image S, a noise image N, and a corrupted image $S \cup N$ formed by the union of S with N. The filter is spatially (translationally) invariant, and because of increasing monotonicity and antiextensivity,

$$S \circ B \subset (S \cup N) \circ B \subset S \cup N \tag{3.14}$$

so that the filtered image lies between the opened uncorrupted image and the noisy image. In Fig. 3.3 the filter has performed well because the structuring element has passed most of the image, while passing little noise. Figure 3.6 illustrates the situation where $(S \cup N) \circ B = S \circ B$.

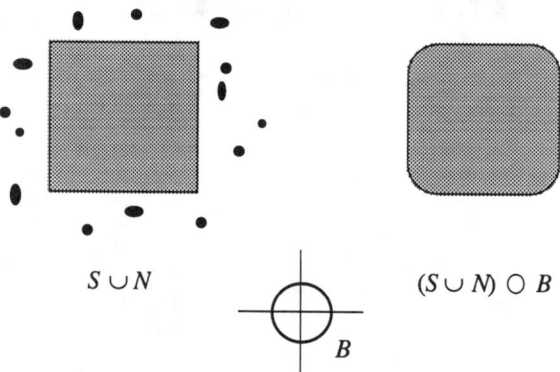

$S \cup N$ B $(S \cup N) \bigcirc B$

Fig. 3.6. Filtering by opening.

The filter effect becomes clearer if we consider the uncorrupted grain-type image S of Fig 3.7, the noise image N of Fig 3.8, and the corrupted image $S \cup N$ of Fig. 3.9. Here, the largest noise grain is smaller than the smallest uncorrupted-image grain and the union is constrained so there is no signal-noise overlapping. Opening with a disk B whose radius is between that of the largest noise grain and the smallest uncorrupted-image grain will yield perfect restoration. Such would likely not be the case if the noise image were to overlap the uncorrupted image: however, if overlapping is modest, then restoration will be close to perfect.

The situation is complicated if some noise-grain radii exceed some uncorrupted-image-grain radii, such as with the uncorrupted, noise, and corrupted images of Figs. 3.10, 3.11, and 3.12, respectively. The open-filtered image is shown in Fig. 3.13, where the disk we have chosen to open by has yielded a fairly good restoration of the uncorrupted image; however, restoration is not perfect. In fact, although we will not go into detail, the structuring-element radius has been chosen so as to optimally restore S from the degraded image $S \cup N$, that optimization resulting from treating the image and noise as random processes and proceeding with a statistical optimization analysis [40].

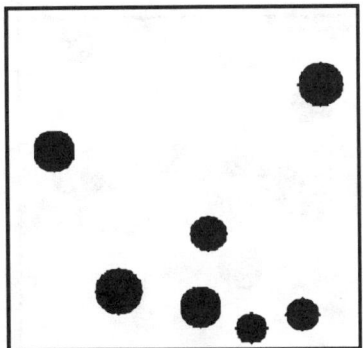

Fig. 3.7. Uncorrupted grain image.

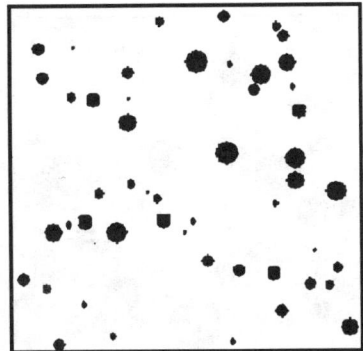

Fig. 3.8. Noise image.

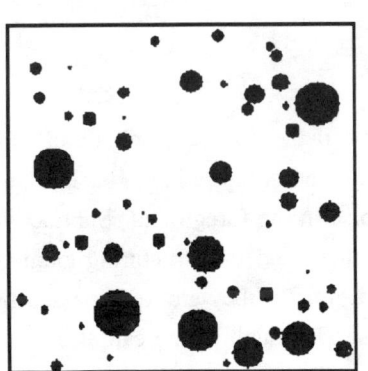

Fig. 3.9. Corrupted image.

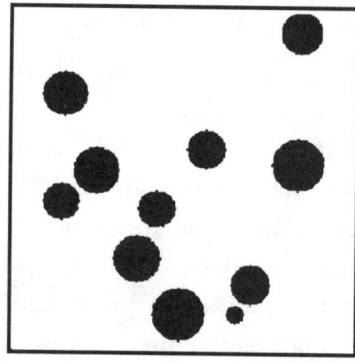

Fig. 3.10. Uncorrupted image.

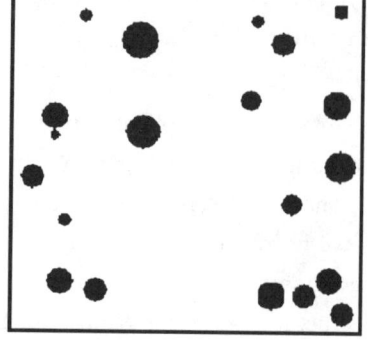

Fig. 3.11. Noise image.

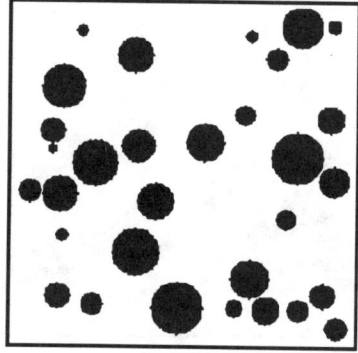

Fig. 3.12. Corrupted image.

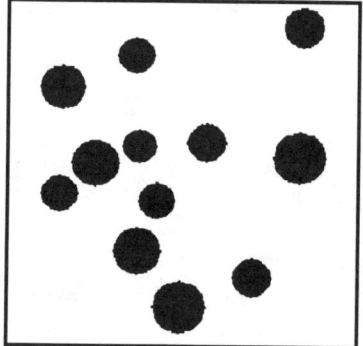

Fig. 3.13. Filtered image.

3.3. Alternating Sequential Filters

Filtering by openings and closings is more problematic if there is both union and sub-tractive noise. An obvious strategy is to open to eliminate union noise in the background and then to follow with a closing to fill subtractive noise in the foreground. Subtractive noise is filled because it is part of the image complement and the structuring element does not fit into those portions of the complement, as it must if they are to be passed by the closing. The resulting filter is called an **open-close**. By duality, one can also close and then open, the filter then being called a **close-open**.

As an illustration, consider the text image of Fig. 3.1 and the salt-and-pepper degradation of Fig. 3.14. The open-close with a 2×2 structuring element is shown in Fig. 3.15.

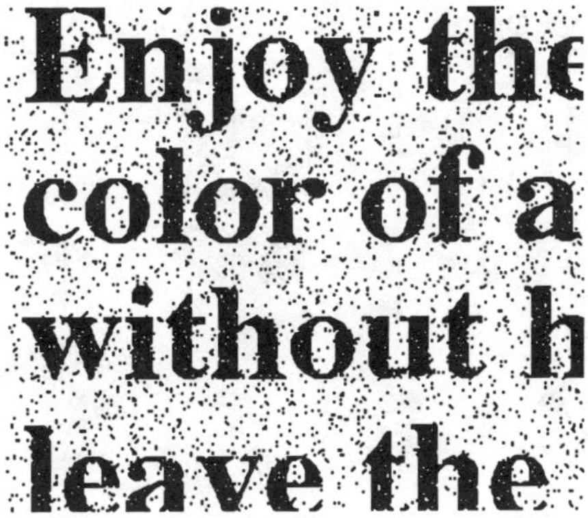

Fig. 3.14. Text image degraded by salt-and-pepper noise.

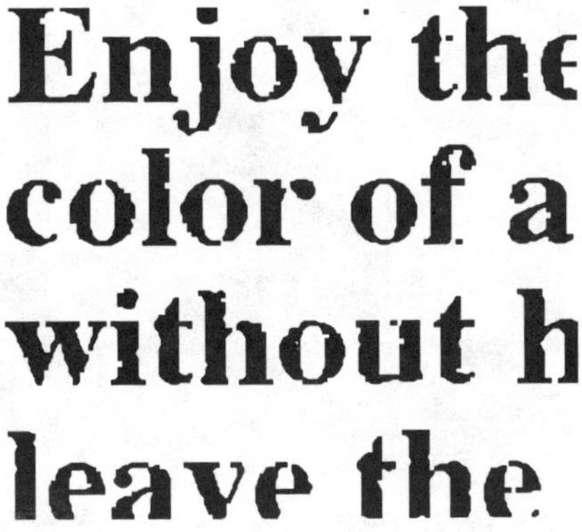

Fig. 3.15. Open-closed salt-and-pepper degraded text image.

A potential pitfall of the open-close strategy occurs when large noise components need be eliminated but a direct attempt to do so will destroy too much of the original image. This problem is illustrated in Fig. 3.16, where an originally perfect rectangle has been degraded by both union and subtractive noise. If we assume the large grain in the lower left is noise, then the open-close has passed that noise grain. In an effort to produce an open-close filter that will eliminate that unwanted grain, we might try to employ a larger disk. Unfortunately, such an attempt can backfire because, as depicted in Fig. 3.17, the disk cannot fit between the salt grains during the initial opening, and the resulting open-close yields an image in which the rectangle has been severely degraded.

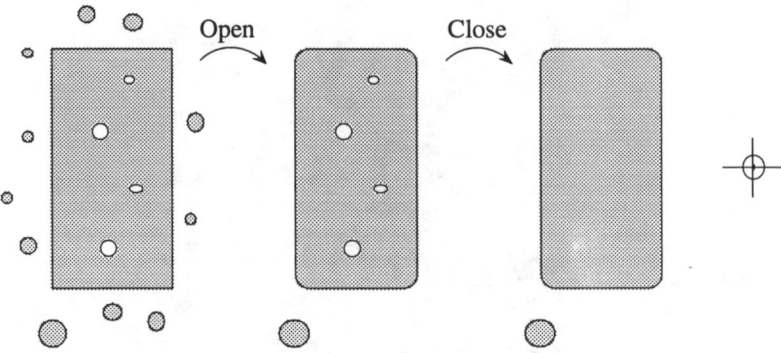

Fig. 3.16. Open-close with small disk.

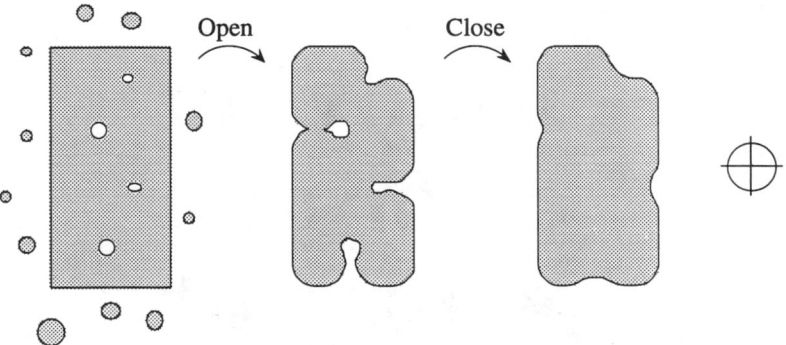

Fig. 3.17. Open-close with large disk.

One way around the dilemma is to employ an **alternating sequential filter**, or **ASF** [172,197]. Here, open-close (or close-open) filters are performed iteratively, beginning with a very small structuring element and then proceeding with ever-increasing structuring elements. The strategy is to first eliminate small salt and pepper components, thereby allowing the larger structuring elements to more likely fit when they are eventually applied in the process. Of course, the process must stop at some point or the desired image will be destroyed. ASFs have been extensively studied, including finding optimal stopping sizes for structuring elements [158].

The ASF method is illustrated using the uncorrupted image of Fig. 3.18 and the degraded realization of Fig. 3.19. The images of Figs. 3.20 through 3.23 show the results of ASF filtering using one through four iterated open-closes, respectively, the open-close sequence resulting from the sequence of 2×2, 3×3, 4×4, and 5×5 structuring elements.

Fig. 3.18. Uncorrupted image.

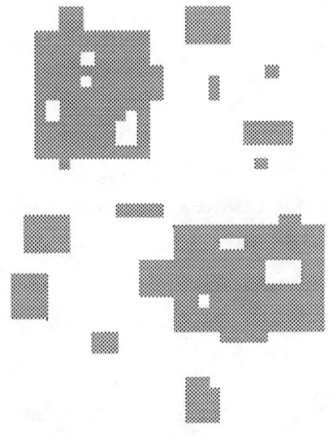

Fig. 3.19. Image corrupted by both union and subtractive noise.

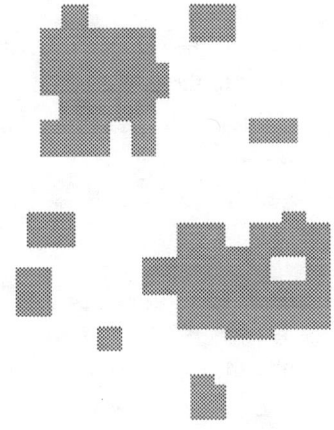

Fig. 3.20. Result of single-stage ASF.

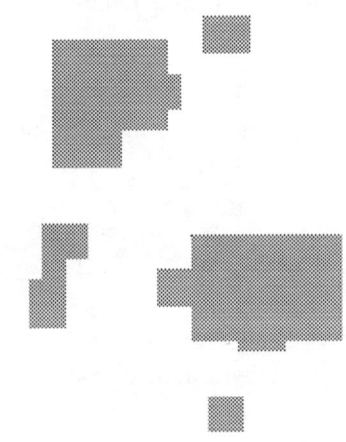

Fig. 3.21. Result of two-stage ASF.

Fig. 3.22. Result of three-stage ASF.

Fig. 3.23. Result of four-stage ASF.

3.4. Invariance

If image A is formed from a union of translations of some shape primitive B, then the opening of A by B yields A, that is, $A \circ B = A$. When $A \circ B = A$, we say that A is **open relative to** B, or that A is **B-open**. From a filtering perspective, A is **invariant** when filtered by opening with structuring element B. A basic proposition states that not only is an image formed as a union of translates of B invariant when opened by B, but unions of translations of B are the only images invariant when opened by B.

A union of translations of an image is a dilation of the image. Thus, B-openness can be reformulated: A is B-open if and only if there exists some image D such that $A = B \oplus D$. A special case of this result occurs when B is a disk of radius r and D is disk of radius s. Then $B \oplus D$ is a disk of radius $r + s$. Hence, a disk is open relative to any disk possessing smaller radius.

Iterated openings by two structuring elements, one open relative to the other, produce the same result as a single opening, the order of iteration being irrelevant: if D is B-open, then

$$(A \circ D) \circ B = (A \circ B) \circ D = A \circ D. \tag{3.15}$$

A key property relative to the analysis of particles and texture regards the effect of opening by different structuring elements when one is open with respect to the other: if D is B-open, then, for any set A, $A \circ D$ is a subset of $A \circ B$. This subset relation can be seen in Fig. 3.24, where D is B-open.

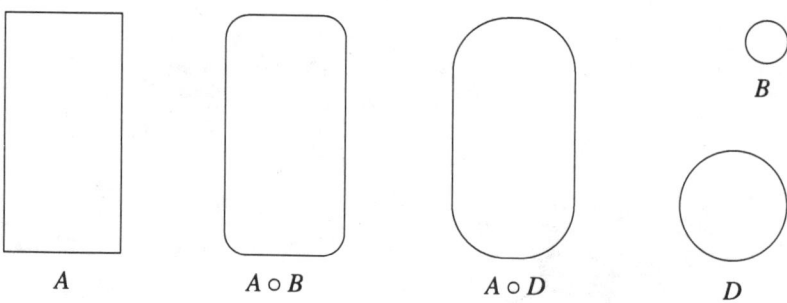

Fig. 3.24. Opening by a relatively open structuring element.

If we consider the image-noise model, $S \cup N$, of Section 3.2, then a special relationship occurs if the uncorrupted image is open relative to the opening structuring element B used for the filter. In such a case, $S \circ B = S$, so that Eq. 3.14 reduces to

$$S \subset (S \cup N) \circ B \subset S \cup N \tag{3.16}$$

so that the filtered image lies between the uncorrupted image and the noisy image.

3.5. Opening by Reconstruction

The corrupted image of Fig. 3.9 is perfectly restored to the ideal image of Fig. 3.7 for several reasons. As noted, there is no intersection between components (disks) and no overlap between the signal-radius and noise-radius distributions. There is, however, a third reason: since the structuring element is a disk of smaller radius than each of the signal components, each signal component is open relative to the structuring element and the signal itself is open relative to the structuring element. Thus, the noise is eliminated and the signal is fully passed. A more general situation arises in the signal-union-noise model if the signal is not open relative to the structuring element.

Suppose the observed image is $S \cup N$, signal and noise components are disjoint, and B is a structuring element such that no translate fits inside a component of N and such that for each component of S there exists a translate of B fitting inside. Then $S \circ B \subset S$ and, if S were B-open, there would be perfect restoration. Under the assumptions, it is possible to obtain perfect restoration by morphological methods by modifying the filter, even when S is not B-open. The problem is to reconstruct the components of S that have been diminished due to opening. Reconstruction is accomplished via conditional dilation.

Before discussing conditioning, we recall the basic connectedness definitions for the Cartesian grid. Two pixels are said to be **strong neighbors** if they are vertically or horizontally adjacent, **weak neighbors** if they are diagonally adjacent, and simply **neighbors** if they are either strong or weak neighbors.

A region within an image (perhaps the entire image) is said to be **strongly connected** if for any two pixels x and y in the region, there exists a sequence of pixels also in the region such that the first pixel is x, the last is y, and each pixel in the sequence is a strong neighbor of the next. A region is simply said to be **connected** if the same definition applies with "neighbor" in place of "strong neighbor." Figures 3.25 (a), (b), and (c) depict strongly connected, connected, and disconnected (not connected) regions, respectively. (A strongly connected region is ipso facto connected.) Every binary image can be expressed as the union of connected regions. If each of these regions is maximally connected, which means that it is not a proper subset of a larger connected region within the image, then the regions are called **connected components** of the image. The image of Fig. 3.25 (c) consists of two connected components. A connected image has one component.

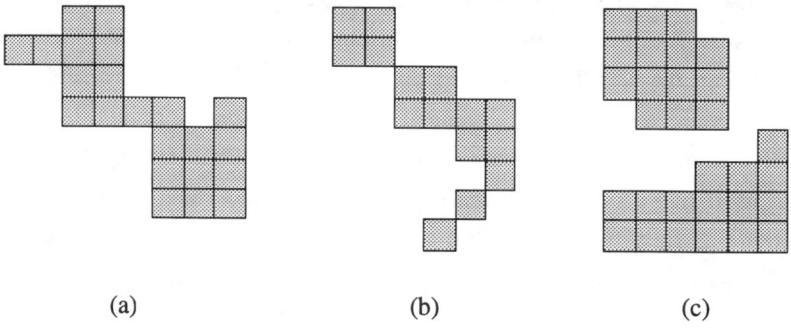

(a)   (b)   (c)

Fig. 3.25. Region connectedness: (a) strongly connected region, (b) connected region, (c) disconnected region.

If an image is dilated by a structuring element containing the origin, it is expanded, and the manner of the expansion depends on the shape of the structuring element. If the dilation is successively repeated, then the original image grows without bound. Growth restriction can be accomplished by **conditioning** the dilation. Referring to the Minkowski addition form of dilation (Eq. 2.16), conditioning is accomplished by restricting in some manner the translates composing the union.

A common form of conditioning (and the only kind considered here) restricts the union-forming translates to a subset of some containing image: if image A is a subimage of C, and B is a structuring element, then the **conditional dilation** of A by B relative to C is defined by restricting the translates to C, the result being

$$A \,(+)\, B = \bigcup \{(B + a) \cap C : a \in A\}, \tag{3.17}$$

where we employ the notation (+) instead of $\oplus$ to indicate there is conditioning. Figure 3.26 illustrates conditional dilation by a disk, part (a) showing the conditioning effect on a single translate and (b) showing the overall conditional dilation. Note that Eq. 3.17 is equivalent to

$$A\,(+)\,B = (A \oplus B) \cap C. \tag{3.18}$$

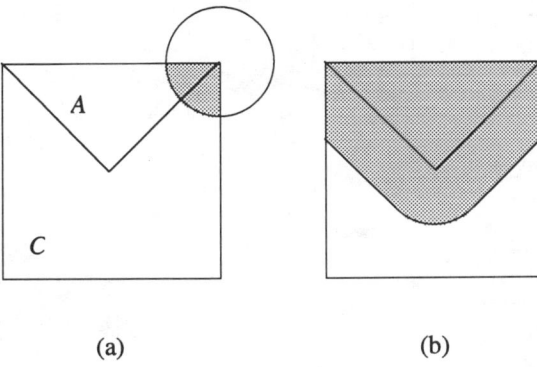

(a)   (b)

Fig. 3.26. Conditional dilation: (a) conditioning of a single translate, (b) output of conditional dilation.

The application of conditional dilation that concerns us here is the reconstruction of an image component containing some **marker**, the latter being a subset of the component. Figure 3.27 shows an image containing three connected components, one of them having a marker pixel. Beginning with the marker, repeated conditional dilations relative to the original image are run with the 3×3 square structuring element, the iteration terminating when no growth occurs. These iterations are depicted in the figure. In the final one, the marked component has been fully filled.

Opening by reconstruction involves opening by a structuring element followed by component extraction using the opened image as marker. The result is a filter that passes unaltered all components of the image into which it is possible to fit a translate of the opening structuring element and eliminates entirely all other components. Figures 3.28 through 3.31 show an uncorrupted signal, a union-noise-degraded realization, the opening of the degraded realization by a 4×4 structuring element, and the opening by reconstruction, respectively.

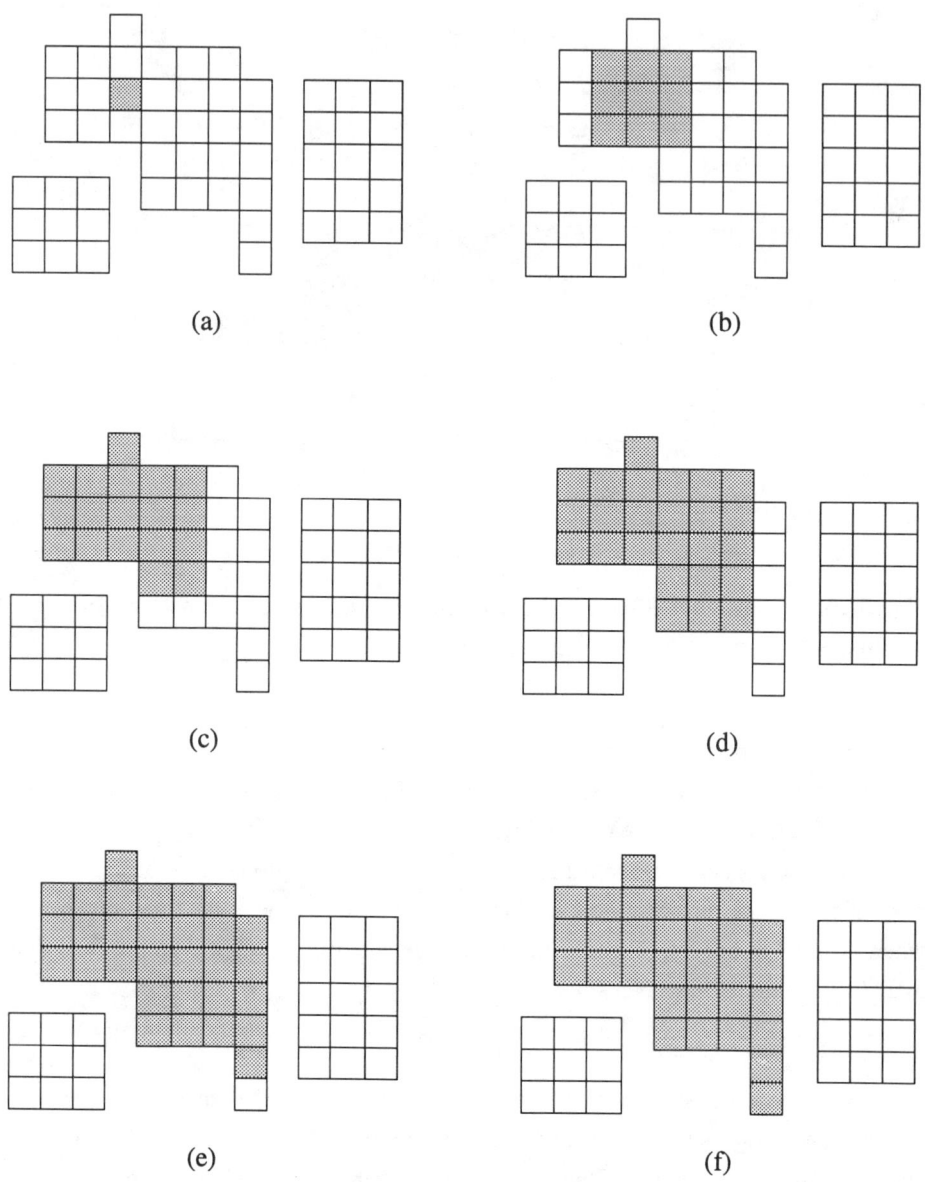

Fig. 3.27. Component extraction.

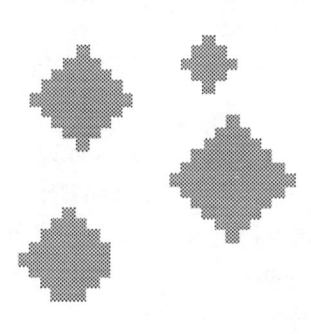

Fig. 3.28. Uncorrupted image.   **Fig. 3.29. Union-noise-corrupted image.**

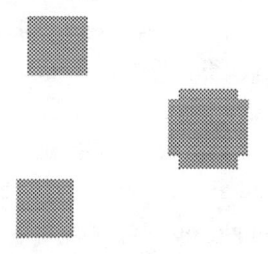

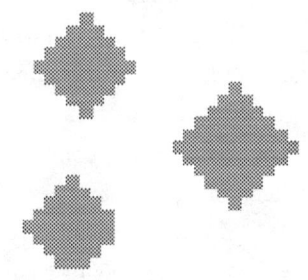

Fig. 3.30. Opened image.   **Fig. 3.31. Opening by reconstruction.**

3.6. τ-Openings

An opening passes only those portions of an image that conform to the shape of the structuring element. Suppose one wishes to pass portions of an image conforming to various primitive shapes, not simply a single primitive. As recognized by Matheron, such an effect can be accomplished by using a filter composed of a number of openings, one for each desirable shape primitive. The final filter output is the union of the individual openings.

Openings satisfy four basic properties: translation invariance, antiextensivity, increasing monotonicity, and idempotence. More generally, any filter satisfying these four properties is called a τ-opening. A fundamental theorem of Matheron [123] states that a filter Ψ is a τ-**opening** if and only if there exists some class **B** of structuring elements such that

$$\Psi(A) = \bigcup \{A \circ B : B \in \mathbf{B}\}. \tag{3.19}$$

B is called a **base** for Ψ. A base is not unique: different bases can produce the same filter; however, our desire is to use a base with a small number of primitives. Design of a τ-opening requires finding an appropriate base. From an algebraic perspective, the four properties, translation invariance, antiextensivity, increasing montonicity, and idempotence, characterize a class of filters (those filters satisfying the four properties), and Matheron showed this class to be composed precisely of unions of openings.

Key to understanding the action of a filter is its **invariant class**, which for a filter Ψ consists of all images that are invariant under Ψ, namely, those images A for which $\Psi(A) = A$. Such images are called **invariants** or **fixed points** of Ψ. If possible, we would like a filter to possess as invariants those images that are considered to be uncorrupted, while at the same time removing noise from corrupted images. τ-openings are particularly straightforward to deal with when it comes to invariance: the invariant class of a τ-opening consists precisely of those images that are formed as unions of translates of base primitives. This is in complete accord with the situation for a single opening, for we have already noted that the invariants of a single opening are those images that are unions of translates of the opening structuring element.

For design purposes in the presence of union noise, if we can express an image as a union of desirable primitives, then we can construct a τ-opening that passes the image in its entirety. Of course, should the noise also be in part made up of some of the same primitives, then some of it will also be passed. Just as in the case of linear filtering, τ-opening filtering requires a trade-off: it is usually necessary to filter out some of the image and to pass some of the noise in order to obtain an optimally filtered image. Such an optimization analysis requires the image and noise to be treated as random processes [40], and we will not examine the issue here; instead we will confine ourselves to a practical example in which the base primitives have been selected in accordance with image geometry.

Figure 3.32 shows a micrograph of silver halide T-grain crystals in emulsion. An algorithm has been developed to segment and geometrically characterize the crystals; however, here we will only focus on a small piece of the algorithm. After some edge processing, the binary edge image of Fig. 3.33 results. In the image we essentially see three types of objects: well-formed crystals possessing definitive geometric shape, "blobs" that are fairly large but are not well-grown crystals, and small ill-formed background "noise" that could have occurred for various reasons. We would like to remove the background noise while preserving the crystals, the effect on blobs not being important and the preservation of crystal geometry being paramount. We recognize that crystals are composed of small line segments and corners. We therefore choose the following τ-opening base:

$$\begin{pmatrix} 0 & 0 & 0 \\ 1 & 1 & 1 \\ 0 & 0 & 0 \end{pmatrix} \quad \begin{pmatrix} 0 & 1 & 0 \\ 0 & 1 & 0 \\ 0 & 1 & 0 \end{pmatrix} \quad \begin{pmatrix} 1 & 0 & 0 \\ 0 & 1 & 0 \\ 0 & 0 & 1 \end{pmatrix} \quad \begin{pmatrix} 0 & 0 & 1 \\ 0 & 1 & 0 \\ 1 & 0 & 0 \end{pmatrix}$$

$$\begin{pmatrix} 0 & 1 & 0 \\ 0 & 1 & 1 \\ 0 & 0 & 0 \end{pmatrix} \quad \begin{pmatrix} 0 & 1 & 0 \\ 1 & 1 & 0 \\ 0 & 0 & 0 \end{pmatrix} \quad \begin{pmatrix} 0 & 0 & 0 \\ 1 & 1 & 0 \\ 0 & 1 & 0 \end{pmatrix} \quad \begin{pmatrix} 0 & 0 & 0 \\ 0 & 1 & 1 \\ 0 & 1 & 0 \end{pmatrix}.$$

Since the origin position plays no role in opening, it need not be indicated when listing base primitives. The result of applying the τ-opening generated by this base is shown in Fig. 3.34. As desired, background noise is virtually eliminated. Had we chosen the primitives to be a bit larger, there would have been more noise filtering; however, desired crystal portions might also have been eliminated, particularly because we have been limited to vertical, horizontal, and diagonal structuring elements, whereas crystal orientations are random.

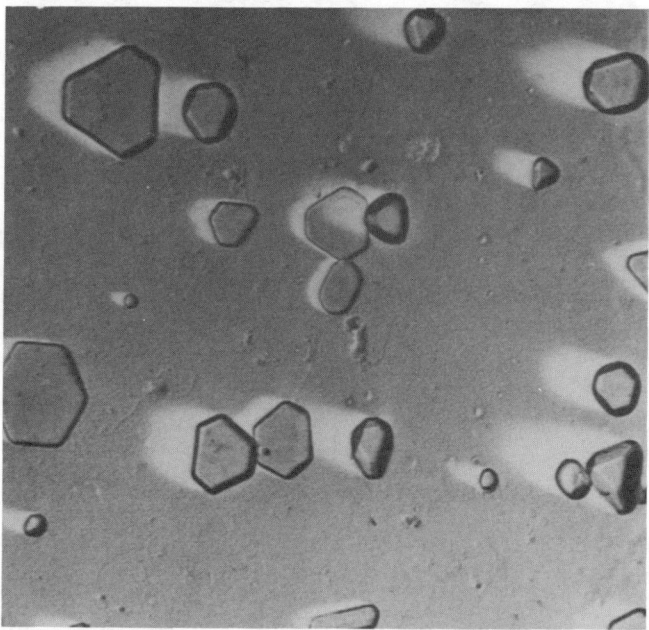

Fig. 3.32. Silver halide T-grain crystals.

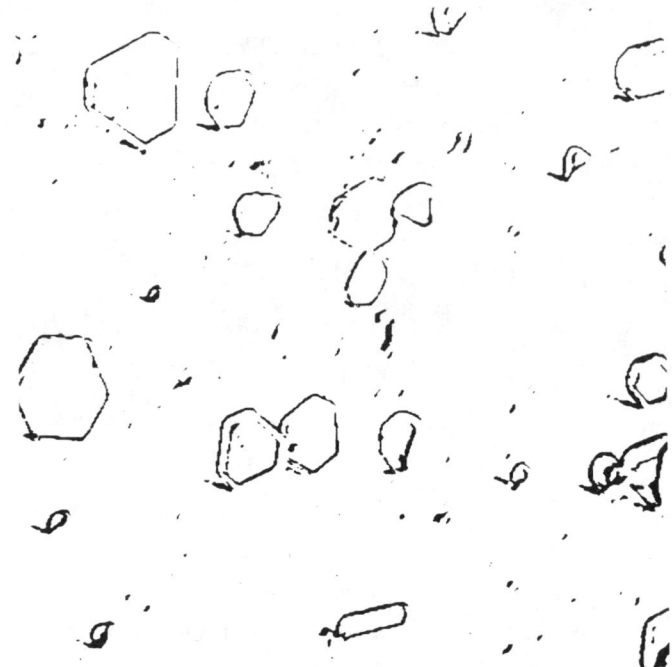

Fig. 3.33. Edge image from crystals.

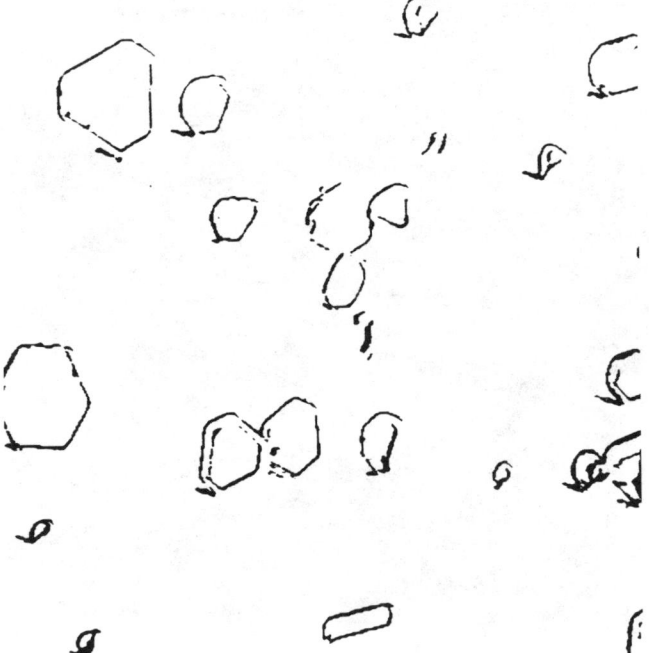

Fig. 3.34. τ-opened edge image.

Dual comments apply to closings. A τ-**closing** is a filter that is increasing, idempotent, extensive, and translation invariant. A filter is a τ-closing if and only if it is an intersection of closings. Whereas τ-openings are good for filtering union noise corrupting the background, τ-closings perform in a dual manner and are good for filtering subtractive noise corrupting the foreground. Design strategies are analogous.

3.7. Granulometries

The present section provides a brief introduction to the granulometric method developed by Matheron for size and shape analysis of granular images [123]. The method is defined via parametric openings and represents a powerful nonlinear method for texture-based classification and detection [19,50,51] and has been employed for shape analysis [117]. It characterizes granular images by means of the manner in which they are sieved through various sized and shaped sieves.

Imagine a sieve through which one might perhaps pan gold. If an image is considered as a collection of grains, then whether or not an individual grain passes through the sieve depends on its size and shape relative to the mesh of the sieve. By increasing mesh size, while keeping the basic mesh shape, more and more of the image will pass through, the eventual result being that no grains will remain. Of course, this sieving model does not fully describe even a granular image, for in a real image grains will likely overlap; nevertheless, it does serve to characterize the removal of nonconforming image structure, and can be developed to obtain image signatures based on the rate of sieving. Granulometric analysis is based on this sieving model.

We consider a basic type of granulometry for Euclidean images. If B is convex and $r > s > 0$, then rB is sB-open, which means that $rB \circ sB = rB$, where $rB = \{rb : b \in B\}$ is the scalar multiple of the set B by r. Consequently, $A \circ rB$ is a subset of $A \circ sB$. If we think of the image A falling through the holes sB and rB, more will fall through the hole rB, thereby yielding a more diminished filtered image. Indeed, since rB is sB-open, filtering by both, in either order, simply yields $A \circ rB$.

If we consider $t > 0$ as a variable, the class of images $\{A \circ tB\}$ is called a **granulometry**, and the primitive B is said to be a **generator** of the granulometry. If $\Omega(t)$ is the area of $A \circ tB$, with $\Omega(0)$ being the area of A itself, then $\Omega(t)$ is a decreasing function of t. Under the assumption that A is not of infinite extent (which is certainly reasonable for image processing), $\Omega(t) = 0$ for sufficiently large t. $\Omega(t)$ is called a **size distribution**.

A normalized size distribution is defined by

$$\Phi(t) = 1 - \frac{\Omega(t)}{\Omega(0)}. \tag{3.20}$$

Since $\Phi(t)$ increases from 0 to 1 and is continuous from the left [123], it is a probability distribution function and its derivative $d\Phi(t) = d\Phi(t)/dt$ is a probability density. Both Φ and $d\Phi$ are known as **granulometric size distributions**. More recently they have come to be known as either the **pattern spectrum** or **granulometric spectrum** of the image relative to the granulometry (or, relative to the generator).

For practical application, the granulometric method must be adapted to digital images. This cannot be done directly owing to two difficulties regarding the Cartesian grid: first, the lack of an appropriate notion of convexity; second, the inability to apply scalar multiplication by arbitrary real numbers. In the second instance, even if we restrict ourselves to integer scalar multiplication, there are still difficulties, since digital images without holes may have holes after scalar multiplication.

The granulometric generation method we now discuss is applicable to both Euclidean and discrete images, its main purpose being application to the latter. We consider a sequence $\{E_k\}$, $k = 1, 2, \ldots$, of structuring elements of increasing size, where E_{k+1} is E_k-open for all k, the latter requirement ensuring that $S \circ E_{k+1}$ is a subimage of $S \circ E_k$ for any image S. Opening in turn by the structuring elements yields a decreasing image sequence

$$S \circ E_1 \supset S \circ E_2 \supset S \circ E_3 \supset \ldots \,. \tag{3.21}$$

For each k, let $\Omega(k)$ be the number of pixels in $S \circ E_k$. Then $\Omega(k)$ is a decreasing function of k, (assuming E_1 consists of a single pixel) $\Omega(1)$ gives the original pixel count in S, and (assuming S to be finite) $\Omega(k) = 0$ for sufficiently large k. Applying the normalization of Eq. 3.20 with k in place of t and 1 in place of 0 yields a normalized size distribution $\Phi(k)$. It is a discrete probability distribution function and possesses a discrete derivative

$$d\Phi(k) = \Phi(k + 1) - \Phi(k), \tag{3.22}$$

which is a discrete density (probability mass function). Again, the density is called the granulometric or pattern spectrum. Its moments are employed as image signatures.

We need to form sequences $\{E_k\}$ such that E_{k+1} is E_k-open. Recall that A is B-open if and only if there exists a set D such that $A = D \oplus B$. Thus, one way to produce a sequence of the desired type is to choose a primitive E and let E_1 be a single pixel, $E_2 = E$, $E_3 = E \oplus E$, $E_4 = E \oplus E \oplus E, \ldots$. A second way is to produce them "by hand."

Consider the image of Fig. 3.35(a), in which digital "balls" of four sizes are dispersed about the image. The generating sequence $\{E_k\}$, from which the four balls generating the image are drawn, consists of digital balls of increasing size (the first being a single pixel). As ever-larger balls are employed for the opening structuring elements, the grains (balls) in the image are sieved from the image. As the structuring-element sequence passes each of the four balls that generate the image, translates of the specific structuring element are sieved from the image, the result being the unnormalized size distribution $\Omega(k)$ of Fig. 3.36. Although it is possible (and likely in real images) for the overlapping grains to create larger, irregular compound grains that are not so regularly sieved by the granulometry, such is not the case in our simulated illustration (but it will be in a real-image example to be discussed shortly). The normalized size distribution $\Phi(k)$ and its discrete derivative $d\Phi(k)$ are illustrated in Fig. 3.36. Notice in this simulated example how the pattern spectrum consists of four impulses. These correspond to the four ball sizes and their heights correspond to the relative image area sieved at the four stages of the granulometry in which they are eliminated.

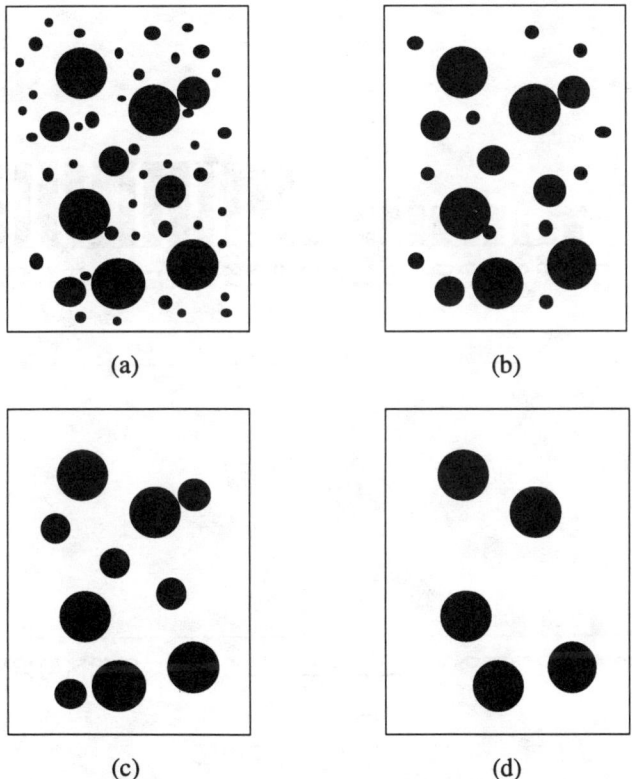

(a)   (b)

(c)   (d)

Fig. 3.35. Ball image and sieving process.

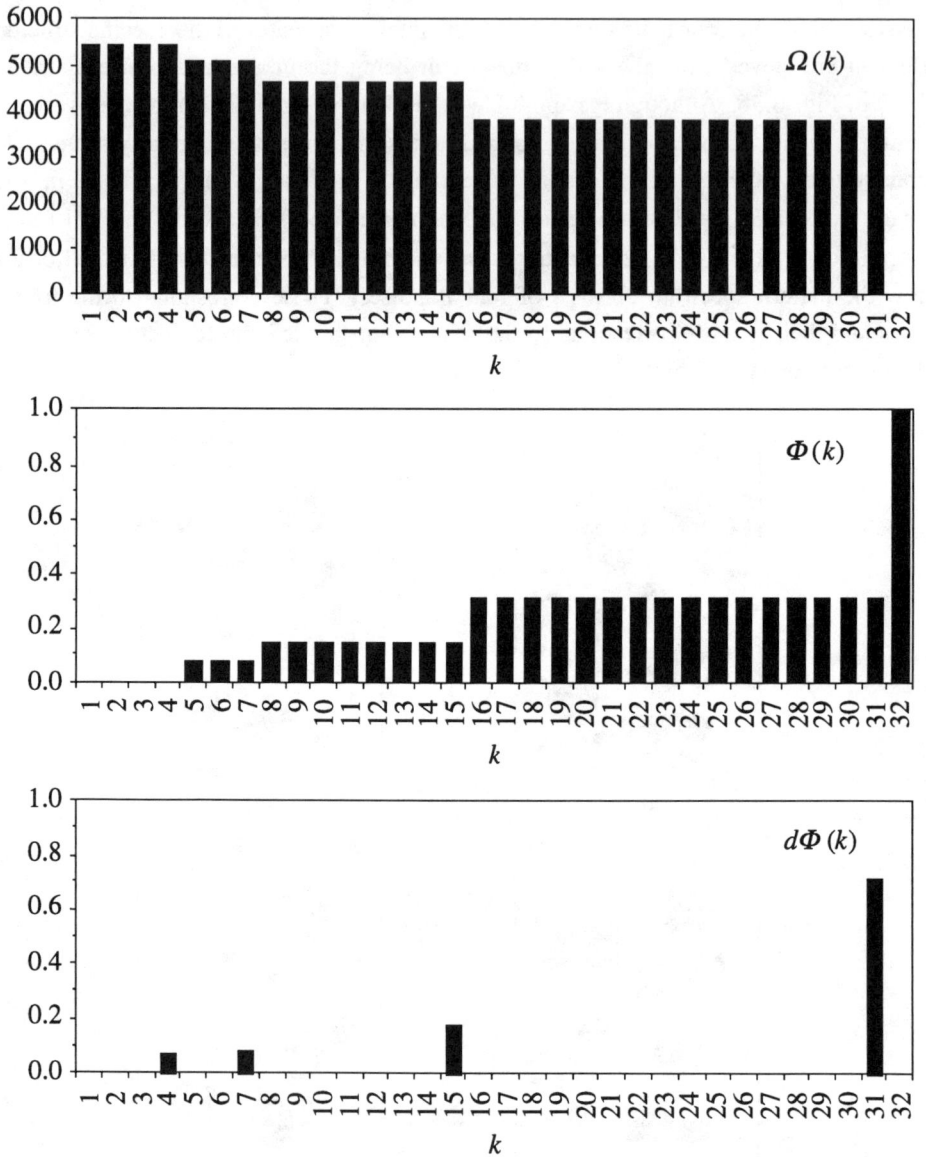

Fig. 3.36. Granulometric size distributions.

We now consider a real-world application to demonstrate the manner in which granulometries can be employed to measure changes in particle-distribution processes. Figures
3.37, 3.38, and 3.39 show toner-particle distributions resulting from an electrophotographic process. In Fig. 3.37, toner particles are spread fairly well uniformly across the
image, whereas in Figs. 3.38 and 3.39 the particles suffer from increasing agglomeration.
Granulometries have been applied to binarized versions of the images using a digital-ball
generating sequence, and the resulting pattern spectra are shown in Fig. 3.40. Notice
how the agglomeration has resulted in a shift of the pattern spectra to the right, especially with regard to skewing. This shifting can result in significant changes in the mean,
variance, and skewness of the pattern spectrum. In fact, hypothesis tests can be based
on these granulometric moments to determine whether, owing to agglomeration (or some
other problem), the electrophotographic process is out of control [52].

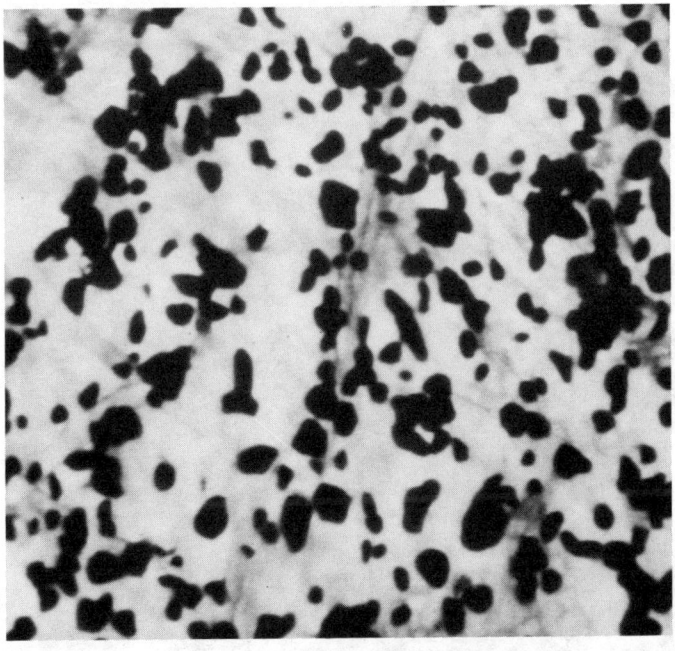

Fig. 3.37. Toner particles with little agglomeration.

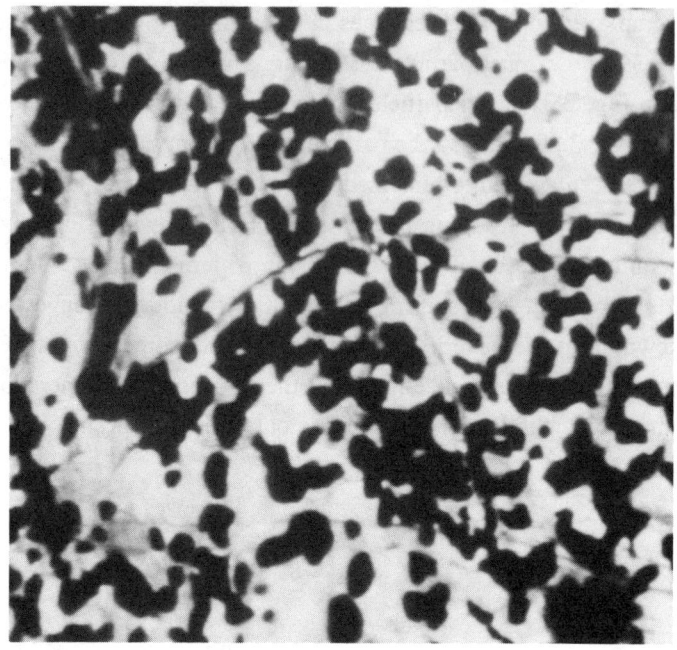

Fig. 3.38. Toner particles with modest agglomeration.

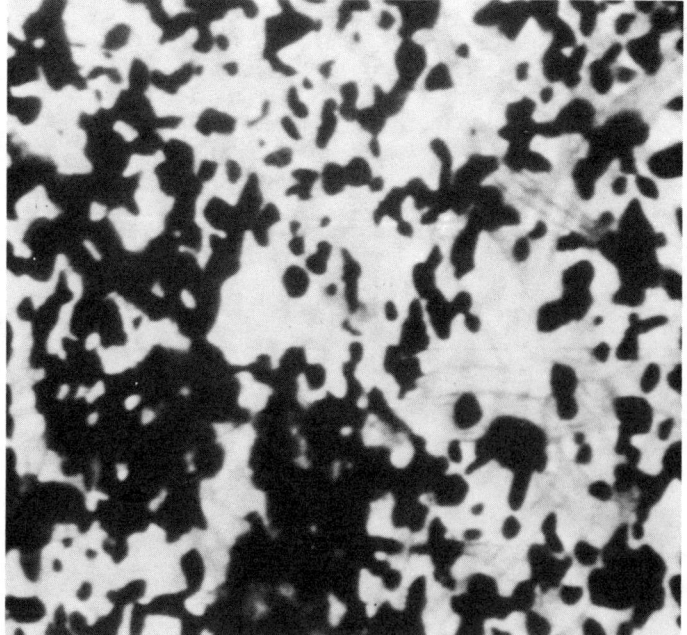

Fig. 3.39. Toner particles with excessive agglomeration.

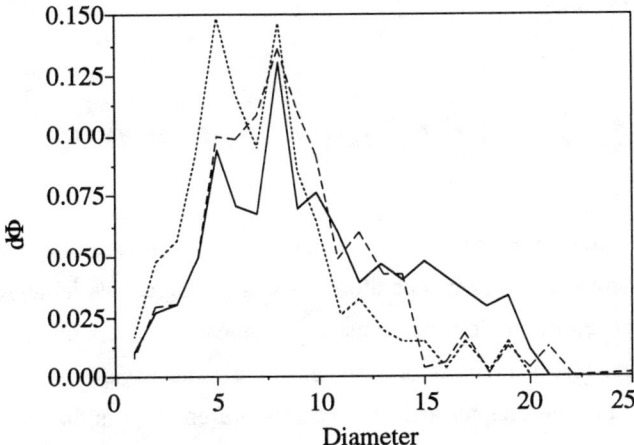

Fig. 3.40. Pattern spectra for toner particles: dotted line for Fig. 3.37, dashed line for Fig. 3.38, continuous line for Fig. 3.39.

Owing to the randomness of the image process, the pattern spectrum is actually a random function (stochastic process): each realization of the image process yields its own particular pattern spectrum, which is a realization of the spectrum process, and each spectrum realization has its own particular moments. Thus, the moments of the pattern spectrum (its mean, variance, skewness, etc.) are themselves random variables. Since these pattern-spectrum moments are random variables, they possess their own statistical distributions, and these in turn possess their own moments. For instance, if we let PSM denote the mean of the pattern spectrum, then PSM is a random variable and it has a mean μ_{PSM} and a standard deviation σ_{PSM}. Classification procedures depend on the statistical distributions (or moments) of the pattern-spectrum moments [11,154,155].

The particular type of granulometry we have discussed is a special, but important, case of the general granulometric theory of Matheron [123] and we leave discussion of the more general theory to the literature [36,63,123,163]. We add only that the theory has been generalized to the gray scale [32] and has been applied to gray-scale texture classification [19,20].

Chapter 4

Representation of Binary Operators

Nonlinear image processing might easily be called "logical image processing," since its roots lie in the theory of Boolean operators on sets of logical variables. In particular, there are important representation theorems for nonlinear operators that are rooted in the fundamental logical representations of Boolean logic. The basic binary representations are covered in the present chapter and gray-scale representation is discussed in the next. Operator representation is not simply a topic of theoretical interest; in fact, it is key to filter design. One need only think of optimal linear filtering, where filter design involves finding optimal weights in the linear representation, to recognize the need for operator representation.

4.1. Window Logic and Increasing Operators

A key motivation for the study of nonlinear digital processing is the logical structure of the computer. Specifically, all algorithms must be reduced to data and instruction flows through logic circuits, and this requirement restricts the kind of mathematical structures that are directly appropriate to digital processing. Linear-space theory is not directly applicable, whereas various nonlinear theories are. The present section focuses on binary digital filtering, in particular, binary operators defined via window (cellular) logic.

A binary **window operator** Ψ is defined via a window W and a Boolean function g on a set $\{x_1, x_2, \ldots, x_n\}$ of binary variables, the functional relation to be written $g(x_1, x_2, \ldots, x_n)$. The window $W = \{w_1, w_2, \ldots, w_n\}$ is a finite (often square) set of pixels and there is a one-to-one correspondence between the variables and the pixels. Ψ is defined at a pixel z by translating the window to z and applying the Boolean function g to the binary values in the translated window (Fig. 4.1). The procedure is in accordance with the digitally stored image representation, 1 if a pixel is in the image and 0 if it is not. If A is the input image, then, geometrically, $z \in A$ if and only if $A(z) = 1$, where $A(z)$ denotes the value of A at z. Because the same Boolean function is applied for each translation of the window, the resulting window operator Ψ is translation-invariant.

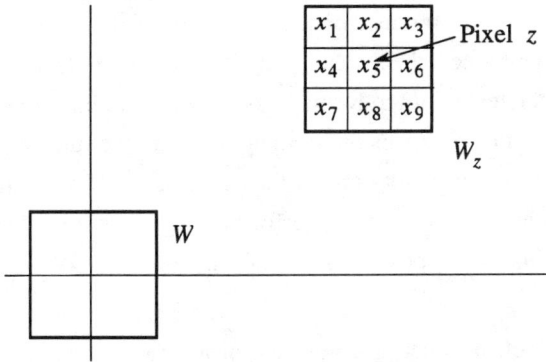

Fig. 4.1. Window logic.

Since g is a Boolean function, it possesses a sum-of-product (maximum-of-minima) representation

$$g(x_1, x_2, \ldots, x_n) = \sum_i x_1^{p(i,1)} x_2^{p(i,2)} \ldots x_n^{p(i,n)}, \tag{4.1}$$

where $p(i,j)$ is either 1 or -1, 1 indicating the presence of the jth variable in the ith product (**minterm**) and -1 indicating the presence of the complemented j^{th} variable in the ith product. There are 2^n minterms forming the union.

Reduction of the minterm expansion can be accomplished in various ways in accordance with the laws of Boolean algebra. The expansion then takes the form

$$g(x_1, x_2, \ldots, x_n) = \sum_i x_{i,1}^{p(i,1)} x_{i,2}^{p(i,2)} \ldots x_{i,n(i)}^{p(i,n(i))}, \tag{4.2}$$

where $x_{i,1}, x_{i,2}, \ldots, x_{i,n(i)}$ denote the $n(i)$ distinct variables in the ith product of the expansion and where there are at most 2^n distinct products in the expansion. For instance, there exist algorithms that provide minimal-gate expansions.

An **increasing Boolean function** is defined just as any other increasing operator: g is increasing if $(x_1, x_2, \ldots, x_n) \leq (y_1, y_2, \ldots, y_n)$ implies $g(x_1, x_2, \ldots, x_n) \leq g(y_1, y_2, \ldots, y_n)$, where $(x_1, x_2, \ldots, x_n) \leq (y_1, y_2, \ldots, y_n)$ if and only if $x_j \leq y_j$ for $j = 1, 2, \ldots, n$. Increasing Boolean functions are usually called **positive Boolean functions**. A Boolean function g is positive if and only if it possesses a sum-of-product representation in which no variables are complemented. Thus, it can be expressed as

$$g(x_1, x_2, \ldots, x_n) = \sum_i x_{i,1} x_{i,2} \ldots x_{i,n(i)}. \tag{4.3}$$

If the set of variables in any product of the expansion contains as a subset the set of variables in a distinct product, then, whenever the former product has value 1, so too does the latter. Thus, inclusion of the former product in the union expansion is redundant and it can be deleted from the expansion without changing the function defined by g. No product whose variable set does not contain the variable set of a distinct product can be deleted without changing the function g. Performing the permitted deletions produces a **minimal representation** of the positive Boolean function.

Consider a positive Boolean function composed of a single product,

$$g_i(x_1, x_2, \ldots, x_n) = x_{i,1} x_{i,2} \ldots x_{i,n(i)}. \tag{4.4}$$

If the window operator Ψ_i is defined via g_i and A is the input image, then $z \in \Psi_i(A)$ if and only if $\Psi_i(A)(z) = 1$. In terms of window logic, this means that when the window W is translated to z, the pixels in the translated window corresponding to the pixels $w_{i,1}, w_{i,2}, \ldots, w_{i,n(i)}$ in W must all have value 1 so that the product of Eq. 4.4 is 1.

Suppose we now view the computation from a different perspective, namely, let B_i denote the set of pixels $w_{i,1}, w_{i,2}, \ldots, w_{i,n(i)}$ in W. Then the pixels in the translated window corresponding to $w_{i,1}, w_{i,2}, \ldots, w_{i,n(i)}$ are $w_{i,1} + z, w_{i,2} + z, \ldots, w_{i,n(i)} + z$, and the product of Eq. 4.4 is 1 if and only if all of these pixel translates lie in the set A, that is, if and only if $B_i + z$ is a subset of A. But this is precisely the definition of erosion. Thus, $\Psi_i(A)(z) = 1$ if and only if $z \in A \ominus B_i$. Put a different way, the single-product operator Ψ_i defined by g_i provides a logical expression of the digital erosion by B_i. This is precisely the reason for the close relationship between cellular logic and binary nonlinear image processing [31,39,198,200].

It is immediate that positive Boolean expressions of the form given by Eq. 4.3 correspond in a one-to-one manner with operators defined by unions of erosions, where the structuring elements corresponding to the erosions are subsets of some predefined pixel window. Letting B_i denote the structuring element corresponding to the product $x_{i,1} x_{i,2}, \ldots x_{i,n(i)}$, the union-of-erosion operator takes the form

$$\Psi(A) = \bigcup_i A \ominus B_i, \tag{4.5}$$

where Ψ is the window operator corresponding to the positive Boolean function defined via Eq. 4.3. Because each positive Boolean expression possesses a minimal form, so too does any union of erosions formed from structuring elements in a finite window. These structuring elements correspond to the minimal products. Taken as a collection, these structuring elements compose the **basis** of the filter and the expansion of Eq. 4.5 can be equivalently taken over the basis.

The identification of Eqs. 4.3 and 4.5 shows the manner in which increasing windowed operators possess isomorphic morphological (erosion) and logical (product) representations. Since practical application often involves windowed operators, the importance of morphological operations for increasing binary filters is clear. Subsequently, this importance for increasing gray-scale filters will also become evident.

4.2. Representation of Increasing Operators

One of the most important theorems of nonlinear image processing is the extension of the classical finite Boolean representation of Eqs. 4.3 and 4.5 to increasing, translation-invariant (not necessarily windowed) operators on subsets of either Euclidean or Cartesian space. We state the Matheron representation and apply it to restoration of text. Relative to restoration, translation invariance means that the restoration filter acts homogeneously across the image and the increasing property means that if one observed image is a subimage of another observed image, then the restored versions maintain the subimage relation.

With every translation-invariant mapping Ψ there is associated a set of images called the **kernel**, written Ker $[\Psi]$. An image lies in the kernel of Ψ if and only if its filtered output contains the origin:

$$\text{Ker}[\Psi] = \{A : 0 \in \Psi(A)\}. \tag{4.6}$$

For instance, suppose Ψ is dilation by a ball of radius r. Then image A lies in the kernel if and only if the dilation of the image by the ball contains the origin.

In its nonreduced form, the **Matheron representation** states that every translation-invariant, increasing operator Ψ can be expressed as the union of erosions by all elements in its kernel:

$$\Psi(A) = \bigcup \{A \ominus B : B \in \text{Ker}[\Psi]\}. \tag{4.7}$$

The representation is redundant since, typically, many of the erosions are unnecessary. The reason is straightforward: if B_1 and B_2 are kernel elements for which B_1 is a subset of B_2, then $A \ominus B_1 \supset A \ominus B_2$, so when taking the union in Eq. 4.7, erosion by

B_2 is unnecessary. This redundancy is exactly analogous to that resulting from having redundant products in the sum-of-product expansion of Eq. 4.3.

A subclass of the kernel is called a **basis** for the operator Ψ if two conditions are satisfied: (1) no image in the basis is a proper subset of another image in the basis; (2) for any image in the kernel, there exists an image in the basis that is a subimage of the kernel image. If a basis exists, it is unique and is denoted by Bas $[\Psi]$. The definition of a basis consists of two parts. The first says that the basis is minimal: there is no redundancy. The second says that it is sufficient for representation: all redundancy has been eliminated but there remain sufficient kernel images so that the representation still applies. It is not our intent here to go into theoretical questions concerning when a basis exists and when it does not. Let it suffice to say that in the digital setting, unless an operator is somewhat pathological, it possesses a basis; indeed, this more general concept of basis is a direct generalization of the basis as a minimal collection of product terms for a positive Boolean function, so that bases always exist for increasing windowed operators. If a basis exists, then the kernel expansion of Eq. 4.7 can be reduced to the equivalent basis expansion [37,63,119,120,121]

$$\Psi(A) = \{A \ominus B : B \in \text{Bas}\,[\Psi]\}. \qquad (4.8)$$

We give two examples of bases in the digital setting. If Ψ is opening by a finite structuring element E, we claim the basis consists of all distinct translates of E containing the origin. Because no translate of E can properly contain a distinct translate of E, the proposed basis lacks redundancy. Second, if a set A lies in the kernel of the opening, then $A \circ E$ must contain the origin, which means there is some translate of E, say $E + x$, such that $0 \in E + x$ and $E + x$ is a subset of A. But by the way we have formulated the basis, the translate $E + x$ lies in it because it contains the origin, so the fact that $E + x$ is a subset of A verifies the second requirement of a basis.

As an illustration, suppose E consists of the strong-neighbor mask (the origin together with its strong neighbors). Then the opening basis consists of five images:

$$\begin{pmatrix} 0 & \mathbf{1} & 0 \\ 1 & 1 & 1 \\ 0 & 1 & 0 \end{pmatrix} \quad \begin{pmatrix} 0 & 1 & 0 \\ \mathbf{1} & 1 & 1 \\ 0 & 1 & 0 \end{pmatrix} \quad \begin{pmatrix} 0 & 1 & 0 \\ 1 & \mathbf{1} & 1 \\ 0 & 1 & 0 \end{pmatrix} \quad \begin{pmatrix} 0 & 1 & 0 \\ 1 & 1 & \mathbf{1} \\ 0 & 1 & 0 \end{pmatrix} \quad \begin{pmatrix} 0 & 1 & 0 \\ 1 & 1 & 1 \\ 0 & \mathbf{1} & 0 \end{pmatrix}. \quad (4.9)$$

According to the Matheron representation, $A \circ E$ can be evaluated by eroding A by the five structuring elements of Eq. 4.9 and then taking the union.

Next consider the binary median Ψ over the strong-neighbor mask. A pixel lies in the filtered image if and only if, among it and its four strong neighbors, at least three pixels lie in the unfiltered image. We show the basis is composed of the following ten elements:

$$
\begin{pmatrix} 0 & 1 & 0 \\ 1 & 1 & 0 \\ 0 & 0 & 0 \end{pmatrix}
\begin{pmatrix} 0 & 1 & 0 \\ 1 & 0 & 1 \\ 0 & 0 & 0 \end{pmatrix}
\begin{pmatrix} 0 & 1 & 0 \\ 0 & 1 & 1 \\ 0 & 0 & 0 \end{pmatrix}
\begin{pmatrix} 0 & 1 & 0 \\ 1 & 0 & 0 \\ 0 & 1 & 0 \end{pmatrix}
\begin{pmatrix} 0 & 1 & 0 \\ 0 & 1 & 0 \\ 0 & 1 & 0 \end{pmatrix}
$$

$$
\begin{pmatrix} 0 & 1 & 0 \\ 0 & 0 & 1 \\ 0 & 1 & 0 \end{pmatrix}
\begin{pmatrix} 0 & 0 & 0 \\ 1 & 1 & 1 \\ 0 & 0 & 0 \end{pmatrix}
\begin{pmatrix} 0 & 0 & 0 \\ 1 & 1 & 0 \\ 0 & 1 & 0 \end{pmatrix}
\begin{pmatrix} 0 & 0 & 0 \\ 1 & 0 & 1 \\ 0 & 1 & 0 \end{pmatrix}
\begin{pmatrix} 0 & 0 & 0 \\ 0 & 1 & 1 \\ 0 & 1 & 0 \end{pmatrix} .
$$

$$(4.10)$$

Direct observation shows that no image in Eq. 4.10 is a proper subset of another. Second, suppose A is in the kernel of Ψ, so that $\Psi(A)$ contains the origin. Then, when the strong-neighbor mask is placed at the origin, at least three of A's pixels in the mask must be 1, so that at least one of the images of Eq. 4.10 is a subimage of A. Hence, the ten images form the basis for Ψ and, according to the Matheron representation, the strong-neighbor median can be evaluated by eroding by each of the ten images and then forming the union.

There is theoretical interest in studying bases of various filters, especially from the standpoint of trying to more fully understand the action of particular filters (for instance, linear filters [42]); however, for practical application, the importance of the Matheron representation lies in a different direction. It tells us that if we wish to construct translation-invariant, increasing filters to restore degraded images, a good approach is to look for bases of structuring elements that will accomplish our desired ends. More specifically, we should look for bases that satisfy certain optimality constraints relative to the type of restoration we wish to accomplish [33,34,111]. This is a topic we will pursue in detail in Chapter 10. For now, we will simply give an example of a restoration filter formed in accordance with the Matheron-representation paradigm.

Consider the 256×256 text image of Fig. 4.2, area coverage being 18.17%. A degraded realization of the image is shown in Fig. 4.3, degradation resulting from subtractive noise. Noise coverage is 10% and each noise component is a randomly generated subset of a 3×3 square. Mean-absolute error for the noisy image is 1.93%. The restored image, shown in Fig. 4.4, is obtained by way of the Matheron expansion using the eight-structuring-element basis of Fig. 4.5, and its mean-absolute error is 0.75%. The basis was found by using the optimization methods of Chapter 10.

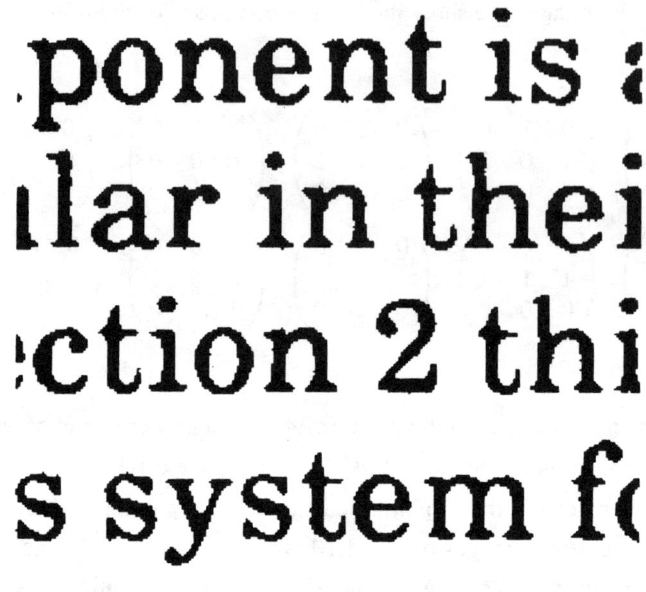

ponent is ɛ

llar in thei

ction 2 thi

s system fc

Fig. 4.2. Text image.

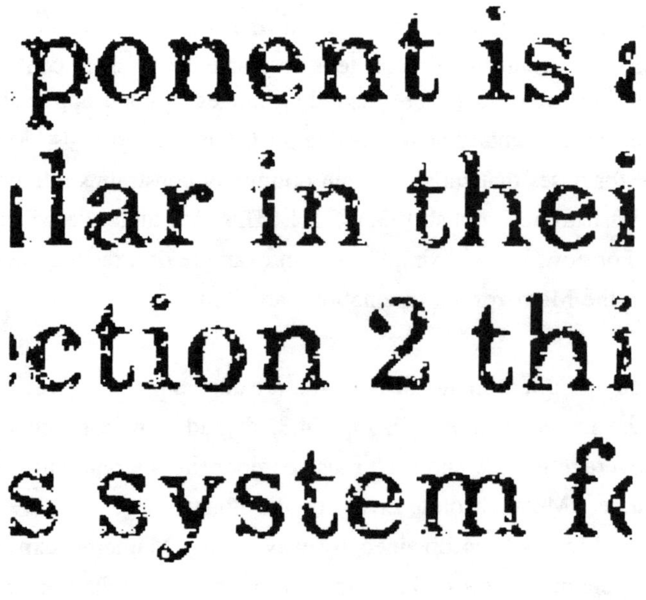

ponent is ɛ

llar in thei

ction 2 thi

s system fc

Fig. 4.3. Noisy text image.

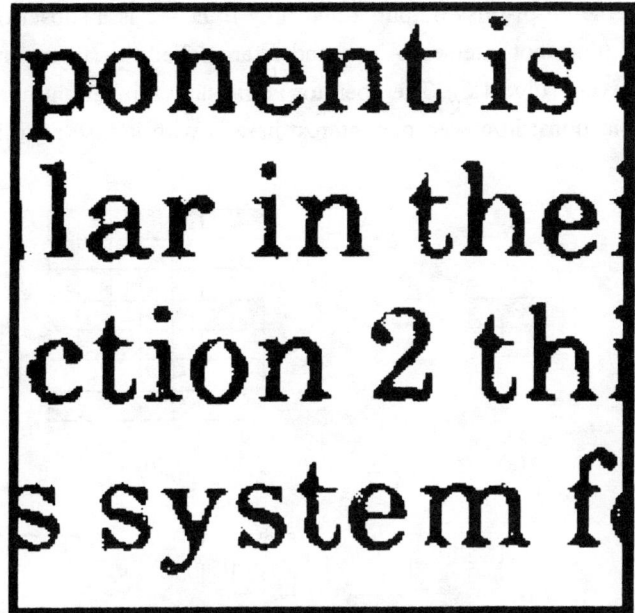

Fig. 4.4. Restored text image.

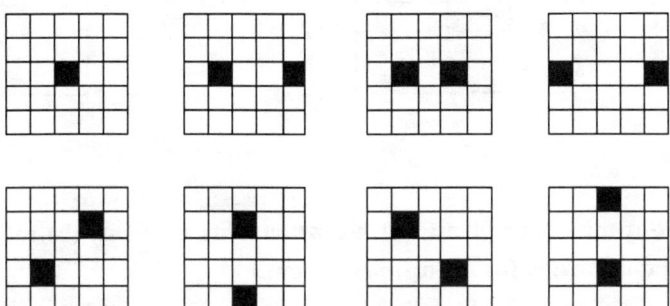

Fig. 4.5. Basis for restoring text image.

4.3. Hit-or-Miss Transform

Erosions provide representation of increasing, translation-invariant operators; a different morphological operation provides representation of operators that are translation-invariant but not necessarily increasing. Given a pair (E, F) of disjoint sets, the **hit-or-miss transform** [163] is defined by

$$A \circledast (E, F) = (A \ominus E) \cap (A^c \ominus F). \qquad (4.11)$$

A pixel z lies in the hit-or-miss output if and only if $E + z$ is a subset of A and $F + z$ is a subset of A^c (does not intersect A). E and F are called the **hit** and **miss** structuring elements, respectively (Fig. 4.6). The operator is translation invariant but nonincreasing. It has many applications; however, our interest here is with its use as a filter.

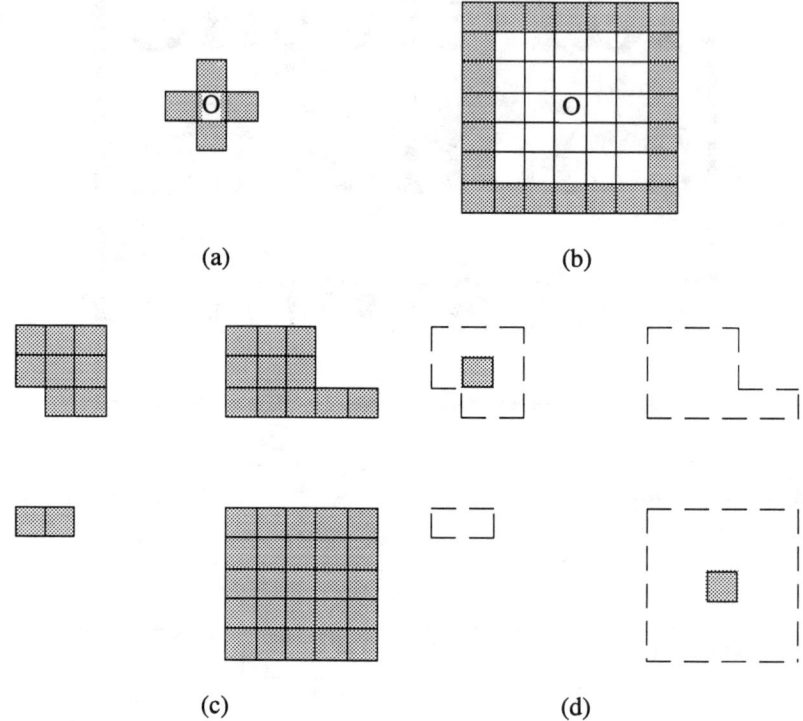

(a)   (b)

(c)   (d)

Fig. 4.6. Hit-or-miss transform: (a) hit structuring element, (b) miss structuring element, (c) input image, (d) hit-or-miss output.

To appreciate the logical genesis of the hit-or-miss transform, consider once again the minterm expansion of Eq. 4.1 for a general Boolean function of n variables. An arbitrary minterm is a single-product Boolean function of the form

$$h_i(x_1, x_2, \ldots, x_n) = x_1^{p(i,1)} x_2^{p(i,2)} \ldots x_n^{p(i,n)}. \tag{4.12}$$

Let E_i and F_i denote the pixel sets in the window W corresonding to the noncomplemented and complemented variables, respectively, in the product. Then $W = E_i \cup F_i$. If Ψ_i is the window operator corresponding to the single-product function h_i, then $\Psi_i(A)(z) = 1$ if and only if $A(w_j + z) = 1$ for all $w_j \in E_i$ and $A(w_k + z) = 0$ for all $w_k \in F_i$. Setwise, $z \in \Psi_i(A)$ if and only if the translate $E_i + z$ is a subset of A and the translate $F_i + z$ is a subset of A^c. In terms of morphological operators, $z \in \Psi_i(A)$ if and only if $z \in (A \ominus E_i) \cap (A^c \ominus F_i)$. Thus,

$$\Psi_i(A) = A \circledast (E_i, F_i). \tag{4.13}$$

Taking the minterm representation of Eq. 4.1 as a whole, it has the equivalent morphological expression

$$\Psi(A) = \bigcup_i A \circledast (E_i, F_i), \tag{4.14}$$

where Ψ is the window operator defined by the Boolean function g of Eq. 4.1 and (E_i, F_i) is the pair of pixel sets corresponding to the ith minterm.

Were we to consider a reduced expression for the Boolean function, say Eq. 4.2, then the same reasoning applies except that W is not necessarily equal to $E_i \cup F_i$. Relative to the window W, a structuring-element pair (E_i, F_i) is said to be **canonical** if $E_i \cup F_i = W$. Relative to a window, we shall denote noncanonical structuring pairs as templates with hit, miss, and don't-care pixels shown black, white, and gray, respectively.

The representation of Eq. 4.14 derives from Boolean algebra; however, just as the finite-window erosion expansion possesses a generalization to increasing, translation-invariant set operators, the finite-window hit-or-miss expansion possesses a generalization to general translation-invariant set operators [9]. There, too, notions of kernel and basis apply; however, one needs to define a concept of minimality, since in logic there does not exist a unique reduction of nonpositive Boolean functions analogous to the unique reduction of positive Boolean functions. We shall not pursue the matter here, contenting ourselves with the existence of representations of the form given in Eq. 4.14 (canonical or otherwise).

4.4. Hit-or-Miss Filtering

Numerous morphological applications depend on the hit-or-miss transform. These include object recognition [22,58,225], thinning (thickening) [163], and pruning. As classically defined, the latter two operations depend on iterative pixel removal via the hit-or-miss transform, not the parallel kind of application given by the representation of Eq. 4.14. In particular, pruning involves the iterative removal of "endpoint" pixels from an image with the goal of removing noise (spurious) pixels. Although our intention is to examine parallel application of the transform, we first describe sequential iterative thinning as it applies to noise removal (there being a dual theory for iterative thickening).

For the structuring pair $B = (E, F)$, the **thinning** of S by B is defined by

$$S \otimes B = S - (S \circledast B). \qquad (4.15)$$

$S \otimes B$ is the set-theoretic difference between S and $S \circledast B$. More generally, a sequence of structuring pairs $B^1, B^2, \ldots$ is employed iteratively to generate a sequence of outputs:

$$\begin{aligned}
S^1 &= S \otimes B^1 \\
S^2 &= S^1 \otimes B^2 \\
S^3 &= S^2 \otimes B^3. \\
&\;\;\vdots
\end{aligned} \qquad (4.16)$$

As the iteration proceeds, the successive sets are ever-thinner.

Classical **pruning** involves iteratively thinning by cycling through a class of structuring elements designed for the purpose of removing spurious pixels. The algorithm is usually limited to one or two cycles; otherwise, it would have a detrimental effect on the underlying image. On each cycle, the algorithm eliminates all endpoint pixels. This is accomplished by cycling through a thinning sequence with the eight structuring pairs of Fig. 4.7, in which the four strong pruners are in the top row and the four weak pruners are in the bottom row.

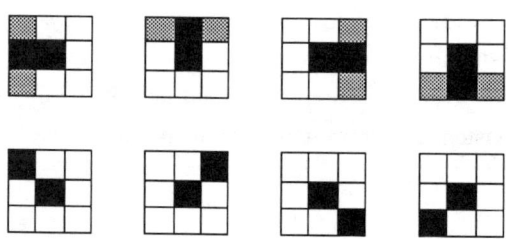

Fig. 4.7. Pruning structuring pairs.

Parallel application involves the expansion of Eq. 4.14. Just as the erosion expansion provides a paradigm for the construction of increasing, translation-invariant filters, the hit-or-miss expansion provides a paradigm for construction of nonincreasing, translation-invariant filters [43]. As in the case of increasing filters, full design consideration involves the theory of optimization, and the nonincreasing-filter optimization theory will be provided in Chapter 10. As in the case of increasing filters, we content ourselves with illustrating the kind of filtering that can be achieved.

If it is assumed that the noisy image results from union noise, then restoration involves removal of pixels; indeed, this is the implicit assumption made with iterative thinning. Thus, we desire an antiextensive filter. A translation-invariant antiextensive filter takes the form

$$\Psi(S) = S - \left[\bigcup_i S \circledast (E_i, F_i) \right].$$
(4.17)

Ψ provides a **parallel thinning**. In effect, the structuring pairs (E_i, F_i) find pixels to be removed from the observed image and the union forms the set of all pixels to be removed. The choice of structuring pairs depends on the type of noise causing the degradation. For the text image of Fig. 4.8 and its union-noise degraded realization of Fig. 4.9, using the eight structuring pairs of Fig. 4.7 in Eq. 4.17 yields the restored image of Fig. 4.10.

Dual to parallel thinning is **parallel thickening**. The filter takes the form

$$\Psi(S) = S \cup \left[\bigcup_i S \circledast (E_i, F_i) \right].$$
(4.18)

Such a filter is translation-invariant and extensive, the pairs (E_i, F_i) locating the pixels to be adjoined to the image.

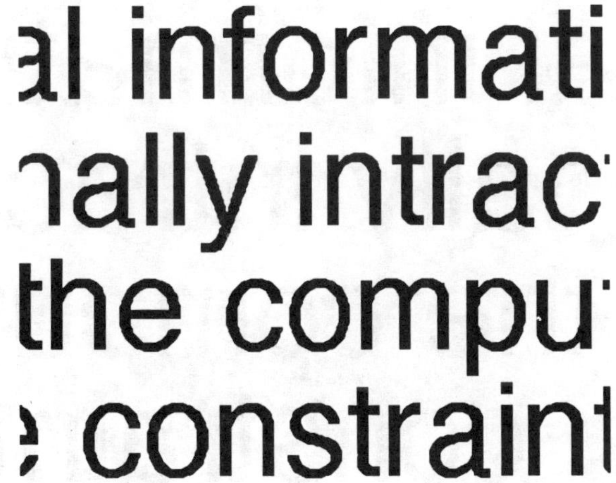

Fig. 4.8. Original text image.

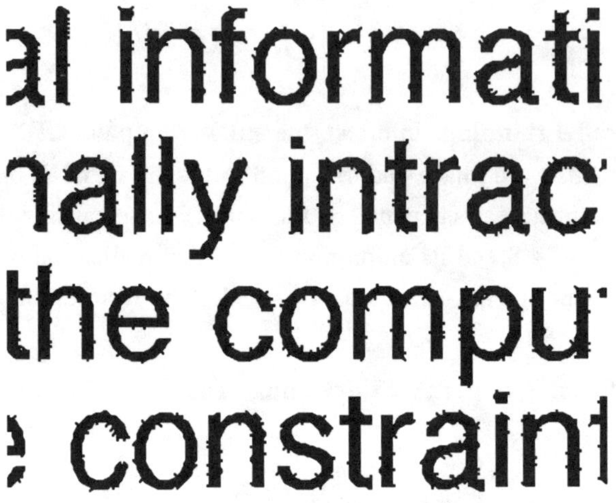

al informati

ally intrac

the compu

constraint

Fig. 4.9. Union-noise degraded image.

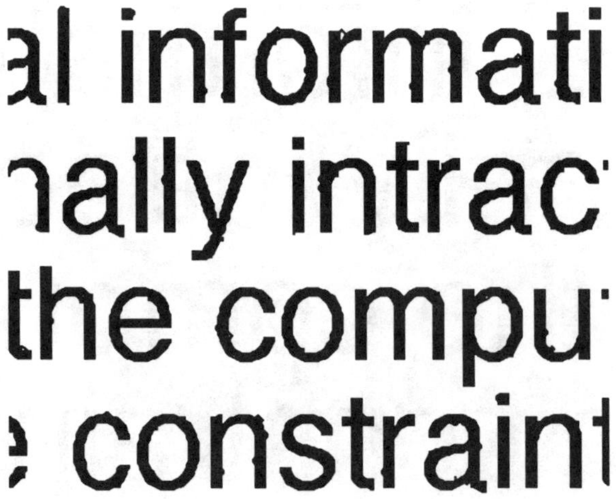

al informati

ally intrac

the compu

constraint

Fig. 4.10. Image filtered by parallel thinning.

Chapter 5

Gray-Scale Morphological Operators

Binary morphological operators play the key role in representing binary filters; gray-scale morphological operators play the key role in representing gray-scale filters. Gray-scale morphological operators act on real-valued functions defined on n-dimensional Euclidean space or the n-dimensional Cartesian grid. For signals, $n = 1$, and for images, $n = 2$. Although our main concern is with image processing, we will develop the gray-scale theory for signals, our aim being to keep notation as simple as possible and to facilitate straightforward illustrative figures. Once the underlying gray-scale theory has been presented for signals, one need only recognize that by treating points on the line as spatial points in the plane the theory at once goes over into the imaging domain, the fundamental point being that the theory is independent of domain dimensionality.

5.1. Mathematical Preliminaries

A Euclidean signal is a real-valued function f defined on a subset of the real line; a digital signal is an integer-valued function defined on a subset of the integers. If the variable is denoted by x, then $f(x)$ denotes the functional value of the signal at x. There are differences between the Euclidean and digital theories; however, to a great extent, these are grist for the theoretician and we do not wish to make them central issues in the present text. Rather, we will develop gray-scale theory from the Euclidean perspective, so as to provide strong geometric intuition, and we will do so making certain implicit assumptions that need not concern the nontheoretician. We are safe on these grounds because, for the digital case, which is the case one would really implement, these implicit assumptions always hold.

Before defining gray-scale erosion, we must provide counterparts for the binary building blocks of translation, subset, union, and intersection. As is typically done, we view a signal in terms of its graph, and the graph can be translated in two ways, spatially or vertically. **Spatial translation** of a signal (image) f by a point (pixel) x is defined by $f_x(z) = f(z - x)$; **vertical translation** of f by y, to be called the **offset** by y, is defined by $(f + y)(z) = f(z) + y$. Applied together, spatial translation and offsetting yield a

translation $f_x + y$, where

$$(f_x + y)(z) = f(z - x) + y. \tag{5.1}$$

Figure 5.1 illustrates a signal f, its spatial translation by x, its offset by y, and the corresponding translation.

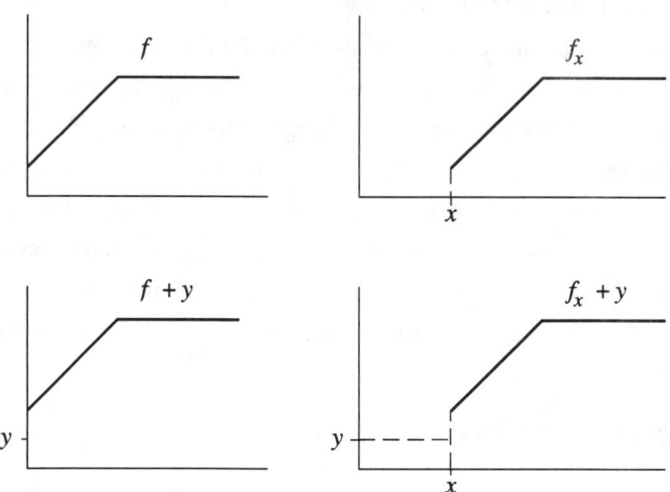

Fig. 5.1. Signal translations.

In the binary setting the notion of subset provides an order relation between images; for the gray scale, the order relation is between signals. In practice, a signal is defined over some domain of finite extent; however, for the purpose of ordering, we assume all signals are defined over the entire real line (Euclidean plane, Cartesian grid), this being accomplished by defining a signal to be $-\infty$ at all points outside its original domain of definition. We also allow a signal to take on the value $+\infty$. We define the **support** of signal f to be

$$D[f] = \{x : f(x) > -\infty\}. \tag{5.2}$$

We say that g is **beneath** f, and write $g \leq f$, if $g(x) \leq f(x)$ for all x. Owing to the minus-infinity convention, $g \leq f$ if and only if $D[g] \subset D[f]$, and for any $x \in D[g]$, which must be the common support of both signals, $g(x) \leq f(x)$. In Fig. 5.2 (a), $g \leq f$; in Fig. 5.2 (b), g is not beneath f because there is a point x in the support of g at which $g(x) > f(x)$; in Fig. 5.2 (c), g is not beneath f because $D[g]$ is not a subset of $D[f]$. For practical application one need not be too concerned with the mathematical subtleties of the negative-infinity convention; however, one must keep in mind the ordering stipulation regarding signal supports.

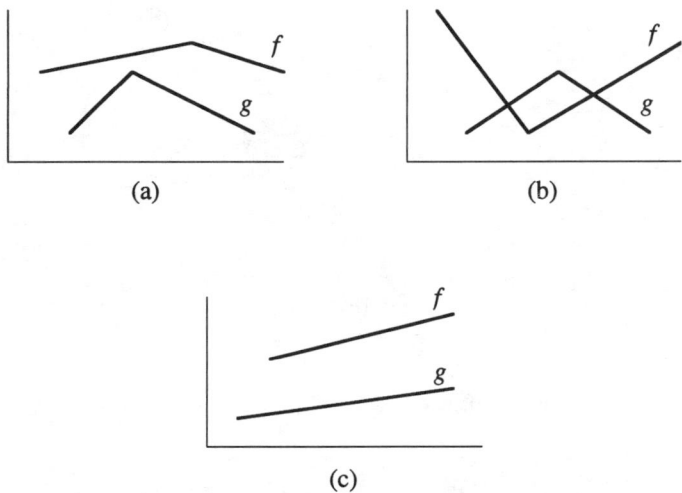

Fig. 5.2. Signal ordering: (a) g beneath f, (b) g not beneath f, (c) g not beneath f.

Corresponding to intersection and union in the binary setting are the minimum and maximum operations between gray-scale signals. The minimum $f \wedge g$ between f and g is defined pointwise by

$$(f \wedge g)(x) = \min\{f(x), g(x)\}, \tag{5.3}$$

where it must be kept in mind that $\min\{a, -\infty\} = -\infty$ for any value a. Relative to supports, $(f \wedge g)(x)$ is the minimum of the two finite values $f(x)$ and $g(x)$ if $x \in D[f] \cap D[g]$; $(f \wedge g)(x) = -\infty$ otherwise. The **maximum** $f \vee g$ is defined pointwise by

$$(f \vee g)(x) = \max\{f(x), g(x)\}, \tag{5.4}$$

where $\max\{a, -\infty\} = a$ for any a. $(f \vee g)(x)$ is the maximum of the finite values $f(x)$ and $g(x)$ if $x \in D[f] \cap D[g]$; $(f \vee g)(x) = f(x)$ if $x \in D[f] - D[g]$; $(f \vee g)(x) = g(x)$ if $x \in D[g] - D[f]$; and $(f \vee g)(x) = -\infty$ if x lies in neither support, that is, if $x \notin D[f] \cup D[g]$. Figure 5.3 shows two signals f and g, their maximum, and their minimum.

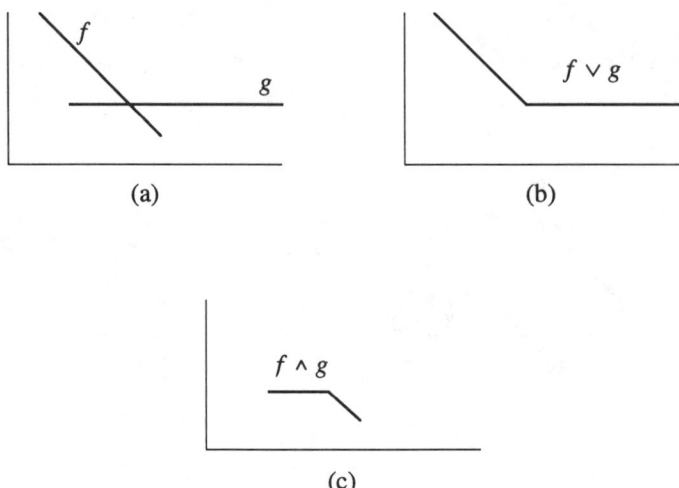

(a)   (b)

(c)

Fig. 5.3. Maximum and minimum between signals: (a) f and g, (b) $f \vee g$, (c) $f \wedge g$.

The foregoing maximum and minimum considerations extend to more than two signals; however, were we to consider an infinite collection of signals, rigorously we would have to replace the maximum and minimum by their continuous counterparts, supremum and infimum. Since our ultimate purpose is digital processing we will forego such an approach, merely accepting the fact that it is possible to make the necessary adjustments to the theory.

We need to consider one more operation, the gray-scale analog to rotation of a set about the origin in the plane. The **reflection** of signal h through the origin is defined by

$$h^{\wedge}(x) = -h(-x). \tag{5.5}$$

Reflection is accomplished by first reflecting the signal through the vertical axis and then through the horizontal axis. It is equivalent to rotating the graph of the signal 180° around the origin.

5.2. Gray-Scale Erosion

Because erosion and dilation satisfy a number of algebraic identities, there are a number of equivalent ways of defining them. We take a fitting approach so as to be consistent with the binary epistemology. The gray-scale **erosion** of signal f by signal (structuri element) g is defined pointwise by

$$(f \ominus g)(x) = \max \{y : g_x + y \le f\}. \tag{5.6}$$

Geometrically, to find the erosion of a signal by a structuring element at a point x, slide the structuring element spatially so that its origin (which for signals is the Euclidean plane origin relative to the structuring element) is located at x and then find the maximum amount the structuring element can be "pushed up" and still be beneath the signal. Since the structuring element must be beneath the signal, x lies in the support of the erosion if and only if the support of the spatially translated structuring element is a subset of the signal support.

Erosions by a semicircular and a flat structuring element are illustrated in Figs. 5.4 and 5.5, respectively. In the first example, the effect is as if the structuring element were "rolled along" under the signal and the origin traced, there being the restriction that the element can never be translated so as not to be beneath the signal. As a disk filters a binary image from the inside, a semicircular element filters the signal from beneath. In both figures, one can observe a fundamental property of gray-scale erosion relative to binary erosion: the support of the gray-scale eroded signal equals the binary erosion of the signal support by the structuring-element support.

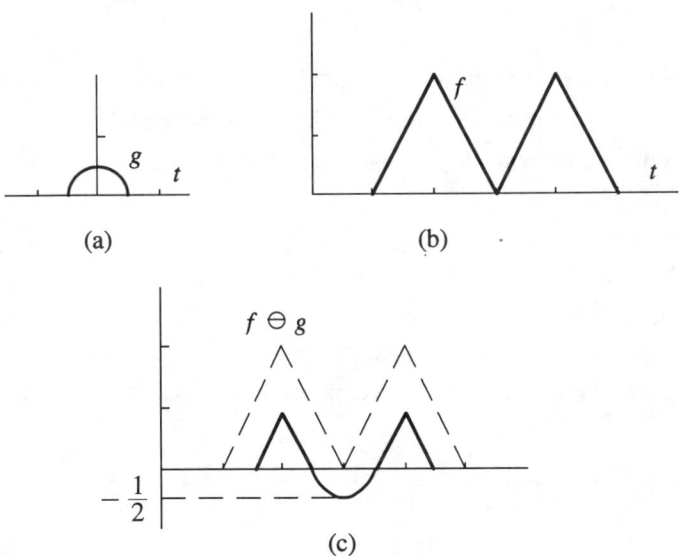

Fig. 5.4. Erosion by a semicircle: (a) structuring element, (b) signal, (c) eroded signal.

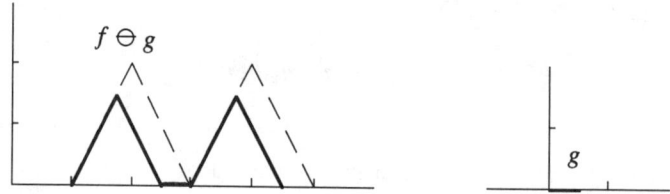

Fig. 5.5. Erosion by a flat structuring element.

Equivalent to the maximum of Eq. 5.6 is the minimum difference between the signal values and the translated-structuring-element values over the support of the translated structuring element:

$$(f \ominus g)(x) = \min \{f(z) - g_x(z) : z \in D[g_x]\}. \tag{5.7}$$

For digital erosion, we consider signals defined on the integers and taking their gray values amongst the integers. A typical signal and structuring element take the forms

$$f = (* \quad * \quad 0 \quad 2 \quad 1 \quad 5 \quad 9 \quad 6 \quad 1 \quad 0) \tag{5.8}$$

$$g = (\mathbf{5} \quad 5 \quad 4) \tag{5.9}$$

respectively, where the asterisk means the signal is $-\infty$ (undefined) at the point and the bold character indicates the origin position relative to the signal. Translating g to the right, the first time it lies beneath f is when it is translated 2 units, so we consider the translate

$$g_2 = (* \quad * \quad 5 \quad 5 \quad 4). \tag{5.10}$$

Applying Eq. 5.7 yields

$$(f \ominus g)(2) = \min \{0 - 5, 2 - 5, 1 - 4\} = -5. \tag{5.11}$$

Successively translating g rightward and taking the minima yields

$$f \ominus g = (* \quad * \quad -5 \quad -4 \quad -4 \quad 0 \quad -3 \quad -4). \tag{5.12}$$

The final point in the support of the erosion is $x = 7$, because beyond that point the translated structuring element no longer lies beneath the signal.

Rather than characterize erosion pointwise, there is a global Minkowski-subtraction formulation:

$$f \ominus g = \bigwedge \{f_x + g^\wedge(x) : x \in D[g^\wedge]\}$$

$$= \bigwedge \{f_{-x} - g(x) : x \in D[g]\}. \quad (5.13)$$

According to this formulation, for each point x in the domain of the structuring element
g, the signal f is translated by $-x$ and $g(x)$ is subtracted from each translated-signal
value. Having one signal for each point in the structuring-element domain, we then take
the pointwise minima of these, keeping in mind the negative-infinity interpretation of the
asterisks.

Applying Eq. 5.13 to the signal and structuring element of Eqs. 5.8 and 5.9 yields three
signals after translation and subtraction:

$$
\begin{aligned}
f_0 - 5 &= (\quad * \quad\quad * \quad -5 \quad -3 \quad -4 \quad 0 \quad\; 4 \quad\; 1 \quad -4 \quad -5 \quad) \\
f_{-1} - 5 &= (\quad * \quad -5 \quad -3 \quad -4 \quad\; 0 \quad\; 4 \quad 1 \quad -4 \quad -5 \quad\; * \quad) \\
f_{-2} - 4 &= (\; -4 \quad -2 \quad -3 \quad\; 1 \quad\; 5 \quad\; 2 \quad -3 \quad -4 \quad\; * \quad\; * \quad).
\end{aligned}
\quad (5.14)
$$

Taking the minima down each "column" of the array in Eq. 5.14 yields the erosion of
Eq. 5.12.

5.3. Gray-Scale Dilation

Dilation can be defined in a dual manner to erosion. Instead of translating the structuring
element and finding the maximum the translated element can be pushed up and still be
beneath the signal, we take the reflection of the structuring element and find the minimum
it needs to be pushed up to be above the signal, when the signal is restricted to the support
of the translated structuring element. This last proviso is necessary because otherwise the
support of the signal would likely extend outside the support of the translated reflected
structuring element and the signal would never lie beneath the translation of the reflected
structuring element. Mathematically, the **dilation** of f by g is defined pointwise by

$$(f \oplus g)(x) = \min \{y : (g^\wedge)_x + y \geq f|D[g^\wedge]\}, \quad (5.15)$$

where $f|D[g^\wedge]$ denotes f restricted to the support of $g^\wedge$. This "fitting" formulation of
dilation is illustrated in Figs. 5.6 and 5.7 for a conical and a flat structuring element,
respectively. Note how the reflection of the conical structuring element is fit into the
signal graph from the top.

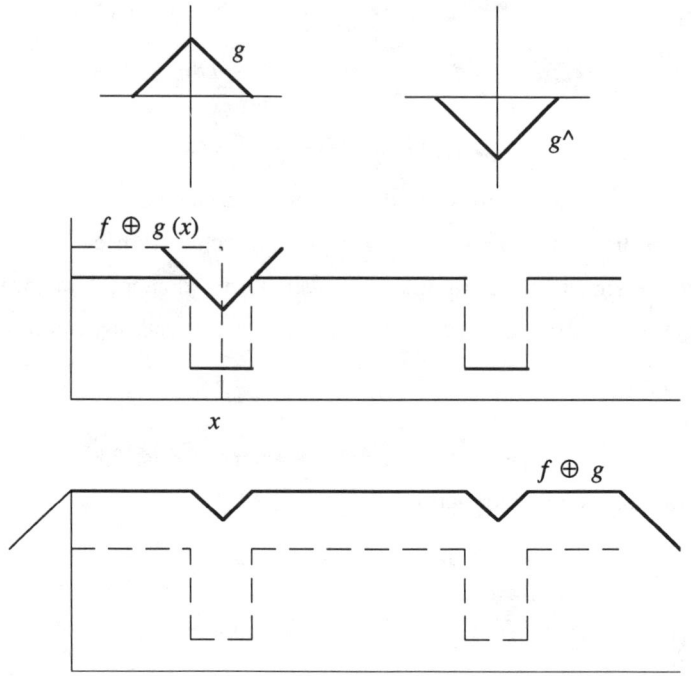

Fig. 5.6. Dilation by conical structuring element.

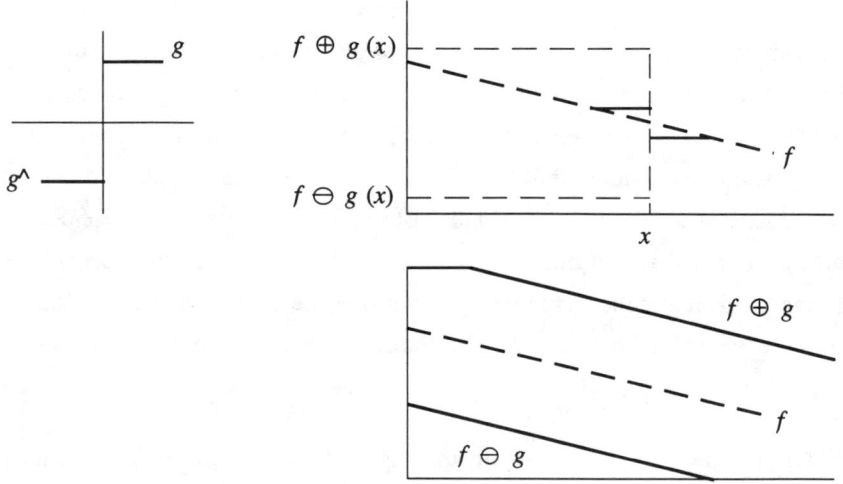

Fig. 5.7. Gray-scale erosion and dilation fittings using a flat structuring element.

There is also a global Minkowski-addition formulation of dilation:

$$f \oplus g = \bigvee \{f_x + g(x) : x \in D[g]\}. \tag{5.16}$$

Here, the signal is spatially translated by each point in the support of the structuring element and offset by the corresponding structuring-element value. Maxima are then taken. Consider the signal and structuring element

$$f = (7 \quad 9 \quad 8 \quad \mathbf{3} \quad 8 \quad 9 \quad 9) \tag{5.17}$$

$$g = (-3 \quad \mathbf{0} \quad -3), \tag{5.18}$$

respectively. Equation 5.16 involves a pointwise maximum of three signals:

$$f_{-1} - 3 = (4 \quad 6 \quad 5 \quad 0 \quad 5 \quad 6 \quad 6 \quad * \quad *)$$

$$f_0 + 0 = (* \quad 7 \quad 9 \quad 8 \quad 3 \quad 8 \quad 9 \quad 9 \quad *) \tag{5.19}$$

$$f_1 - 3 = (* \quad * \quad 4 \quad 6 \quad 5 \quad 0 \quad 5 \quad 6 \quad 6).$$

Taking the maximum down each column of the array yields

$$f \oplus g = (4 \quad 7 \quad 9 \quad 8 \quad \mathbf{5} \quad 8 \quad 9 \quad 9 \quad 6). \tag{5.20}$$

Notice how the support has been expanded, the support of the dilation being the binary dilation of the supports, and notice how the amount of the drop-out at the origin has been attenuated by fitting the "conical" element $g^{\wedge}$ from the top.

As in the binary setting, dilation is both commutative and associative. By commutativity we can interchange the roles of signal and structuring element in Eq. 5.16 to obtain

$$f \oplus g = \bigvee \{g_x + f(x) : x \in D[f]\}. \tag{5.21}$$

As illustrated in Fig. 5.8, in this formulation dilation is accomplished by translating the structuring element so that its origin lies on the signal graph, doing this for each point on the graph, and then taking the maximum over all such copies of the structuring element. This is analogous to the Minkowski-addition formulation of binary dilation, where the structuring element is translated to all points in the binary image and the union taken.

The commutativity of erosion and dilation with intersection and union, respectively, are key properties in the binary setting; in the gray scale, the corresponding properties are commutativity with minimum and maximum, respectively:

$$\left(\bigwedge_i f_i\right) \ominus g = \bigwedge_i f_i \ominus g \tag{5.22}$$

$$\left(\bigvee_i f_i\right) \oplus g = \bigvee_i f_i \oplus g. \tag{5.23}$$

Erosion-dilation duality takes the form

$$f \oplus g = -[(-f) \ominus (-g^\wedge)]. \tag{5.24}$$

In using the duality formula it must be kept in mind that $-(-\infty) = \infty$ and $-(+\infty) = -\infty$. Unless this convention is embedded in software, duality will not hold when digitally processing images.

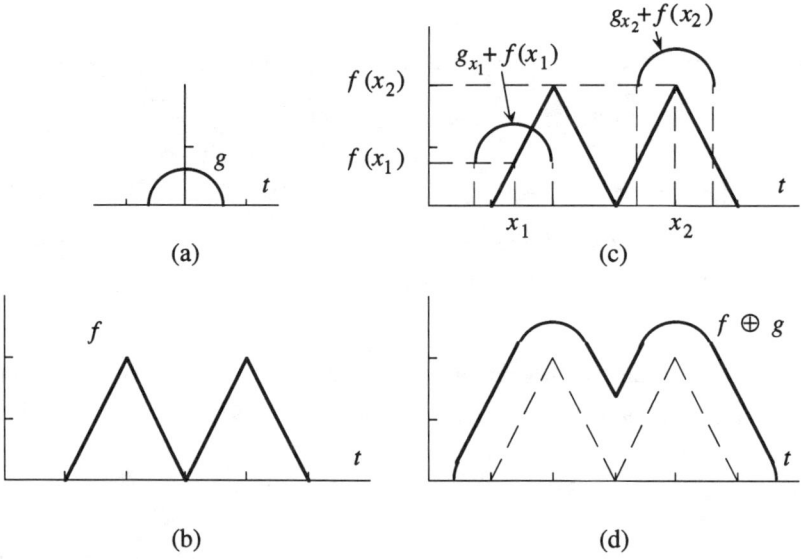

Fig. 5.8. Minkowski-addition formulation of dilation.

5.4. Gray-Scale Opening and Closing

As in the binary case, gray-scale opening and closing can be defined in terms of gray-scale erosion and dilation. Opening and closing are dual and both can be characterized directly in terms of fitting. Gray-scale **opening** is defined by

$$f \circ g = (f \ominus g) \oplus g. \tag{5.25}$$

As in the binary setting it is usually better to view opening in terms of fitting. Here,

$$f \circ g = \bigvee \{g_x + y : g_x + y \leq f\}. \tag{5.26}$$

According to Eq. 5.26, the opening is found by taking the maximum over all translations of the structuring element that fit beneath the input signal. The fitting formulation gives the geometric intuition for opening: slide the structuring element along beneath the signal and at each point record the point on the structuring element translate that is highest at that point. The position of the origin relative to the structuring element is irrelevant.

Gray-scale closing can be defined via duality by

$$f \bullet g = -[(-f) \circ (-g)]. \tag{5.27}$$

Simply flip the signal and structuring element across the horizontal axis, perform the opening, and then reflip. Opening and closing are illustrated in Figs. 5.9 and 5.10, respectively.

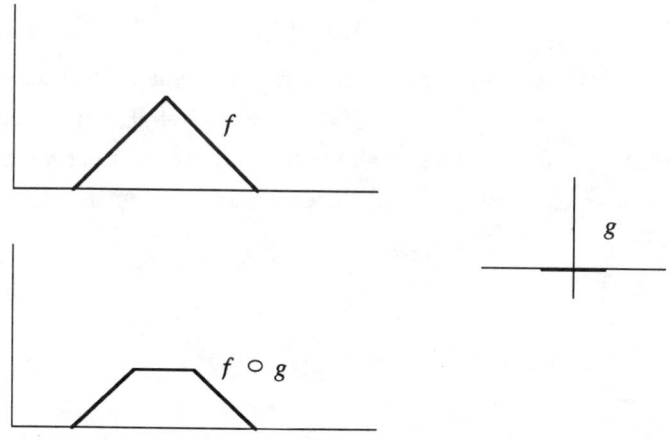

Fig. 5.9. Gray-scale opening.

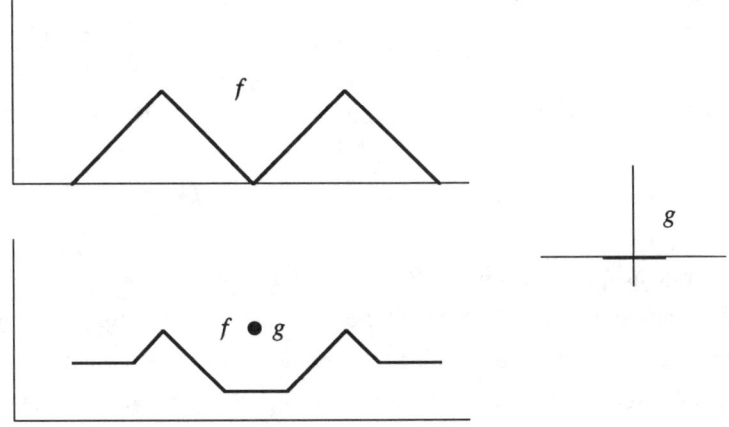

Fig. 5.10. Gray-scale closing.

5.5. Filter Properties

Corresponding to the basic filter properties for binary operators are filter properties for gray-scale operators. A gray-scale filter takes an input signal f and yields an output signal $\Psi(f)$. When treating gray-scale signals (images), it is the topography of the graph as a subset of the plane (space) that is key.

An operator Ψ is said to be **translation invariant** if

$$\Psi(f_x + y) = \Psi(f)_x + y. \tag{5.28}$$

If Ψ is translation invariant, then, ipso facto, it is both spatially translation invariant and offset invariant: $\Psi(f_x) = \Psi(f)_x$ and $\Psi(f + y) = \Psi(f) + y$. On the other hand, it is certainly possible for a filter to be spatially translation invariant but not offset invariant, or vice versa. Gray-scale erosion, dilation, opening, and closing are translation invariant:

$$(f_x + y) \ominus g = (f \ominus g)_x + y \tag{5.29}$$

$$(f_x + y) \oplus g = (f \oplus g)_x + y \tag{5.30}$$

$$(f_x + y) \circ g = (f \circ g)_x + y \tag{5.31}$$

$$(f_x + y) \bullet g = (f \bullet g)_x + y. \tag{5.32}$$

An operator Ψ is **monotonically increasing** if, whenever f is beneath h, then $\Psi(f)$ is beneath $\Psi(h) : f \leq h$ implies $\Psi(f) \leq \Psi(h)$. Erosion, dilation, opening, and closing are increasing: $f \leq h$ implies $f \ominus g \leq h \ominus g$, $f \oplus g \leq h \oplus g$, $f \circ g \leq h \circ g$, and $f \bullet g \leq h \bullet g$.

From the perspective of structuring elements, the order relation is preserved for dilation and inverted for erosion. If $g \leq k$, then $f \oplus g \leq f \oplus k$ and $f \ominus g \geq f \ominus k$. The latter relation is key for filter representation. Its genesis is straightforward. If g is beneath k, then g can be pushed up at least as much as k and still lie beneath f. Consequently, the erosion by g is at least as great as the erosion by k.

An operator Ψ is **antiextensive** [**extensive**] if $\Psi(f) \leq f$ $[f \leq \Psi(f)]$. Gray-scale opening is antiextensive: $f \circ g \leq f$. Closing is extensive: $f \leq f \bullet g$. Ψ is **idempotent** if $\Psi(\Psi(f)) = \Psi(f)$. Opening and closing are idempotent:

$$(f \circ g) \circ g = f \circ g \qquad (5.33)$$

$$(f \bullet g) \bullet g = f \bullet g. \qquad (5.34)$$

A gray-scale filter that is translation invariant, monotonically increasing, antiextensive, and idempotent is called a τ-**opening**. There is a gray-scale extension of Matheron's representation for binary τ-openings [63]: Ψ is a τ-opening if and only if it can be represented as

$$\Psi(f) = \bigvee \{f \circ g : g \in \mathbf{B}\}, \qquad (5.35)$$

where $\mathbf{B}$ is a **base** for Ψ. In the gray scale, the invariant class of Ψ consists of all maxima of translations of signals in $\mathbf{B}$.

The **dual** of filter Ψ is defined by $\Psi^*(f) = -\Psi(-f)$. Equations 5.24 and 5.27 provide the erosion-dilation and opening-closing duality relations, respectively.

As recognized by Serra [166,167], the appropriate abstract mathematical setting for non-linear image processing is lattice theory. The primary reason for this is that lattice theory is concerned with order and, whether we are in the binary setting where set inclusion determines order or in the gray-scale setting where function inequality determines order, except for translation invariance, the basic filter properties (increasing monotonicity, antiextensivity, and idempotence) are defined via the order relation. Indeed, it is for this reason that much of the original binary morphological theory [123] goes directly over to the gray scale once one recognizes that both theories are contained within a more general abstract lattice theory. As for translation invariance, it is a property of the signal domain, whereas the other three properties are range-based; indeed, from a statistical optimization perspective, requiring operator translation invariance is tantamount to an assumption of image stationarity, which is epistemologically far different than the order properties. In any event, nonlinear operator theory, including representation, has been developed independently of translation invariance in the context of lattices [165]. In that context, Serra [169] has defined a **morphological filter** as a lattice operator that is increasing and idempotent. We shall refrain from going into the theory of lattices in the present text, but on occasion will note references when they are directly related to the topic at hand. For instance, the representation of Eq. 5.35 possesses a lattice-theoretic version [169].

5.6. Umbra Transform

There is a close geometric relationship between gray-scale and binary morphology and this relationship can be formalized via the umbra transform [197]. Although the umbra is used very little today, it played a central role in the original development of gray-scale morphology and still provides beneficial intuition.

The graph of a signal f is a subset of the plane and is defined by

$$G[f] = \{(x, f(x)) : x \in D[f]\}. \tag{5.36}$$

The **umbra** of f consists of all points in the plane lying beneath or on the graph of f. It is defined by

$$U[f] = \{(x, y) : x \in D[f] \text{ and } y \leq f(x)\}. \tag{5.37}$$

Given a set A in the plane, we wish to formalize the notion of its "surface." For instance, the surface of a signal's umbra is the graph of the signal. If we assume A is topologically closed, which means it contains its boundary, then defining a surface is straightforward. Rigorously, we define the **surface** of set A by

$$S[A] = \{(x, y) : y \geq z \text{ for any } (x, z) \in A\}, \tag{5.38}$$

where we only consider points x that are first coordinates of points in the set. The surface operator "peels off the top" of a set. We can consider $S[A]$ as either a set in the plane or as defining the graph of a signal. For any signal f, the surface of its umbra is its graph, namely, $S[U[f]] = G[f]$. Thus, as indicated in Fig. 5.11, the surface operator acts as an inverse operator for the umbra transform applied to signals.

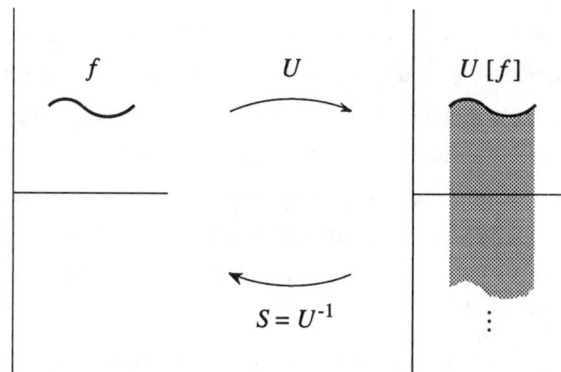

Fig. 5.11. Umbra and surface transforms.

The importance of the umbra in mathematical morphology is that it provides a mechanism for expressing gray-scale operations in terms of binary operations. This correspondence facilitates intuition and can be theoretically useful. There are three fundamental umbra propositions relating binary and gray-scale morphology. For erosion, dilation, and opening, these are

$$f \ominus g = S[U[f] \ominus U[g]] \tag{5.39}$$

$$f \oplus g = S[U[f] \oplus U[g]] \tag{5.40}$$

$$f \circ g = S[U[f] \circ U[g]]. \tag{5.41}$$

In each case, the gray-scale operation can be performed by taking the umbrae of the signal and structuring element, performing the corresponding binary operation using these umbrae, and then taking the surface of the result.

Since we have only defined the surface operator for topologically closed sets, a mathematical question arises as to whether Eqs. 5.39, 5.40, and 5.41 are restricted in any way. In fact, they are not. Although we have restricted our definition to avoid mathematical subtleties, even if the result of the binary operation involving the two umbrae does not contain its boundary, the surface operator can still be given meaning by defining it in a limiting fashion, and the propositions still hold [63].

5.7. Flat Structuring Elements

Flat structuring elements are most important for gray-scale nonlinear image processing; by a flat element we mean one that is zero over its support [163]. The class of flat structuring elements is identical to the set of subsets of the line (plane for images). Consequently, we may consider erosion or dilation by a flat structuring element as erosion or dilation of a signal by a set. Hence, if g is zero on its support D, it is common to write $f \ominus D$ and $f \oplus D$ to denote the erosion and dilation of f by g, respectively. We do not differentiate between erosion by a flat structuring element and erosion by a set. Should a structuring element be constant, but nonzero, over its support, because of translation invariance, we can always offset it, operate by the resulting flat structuring element, and then reoffset in the opposite direction.

Applying the definitions of erosion and dilation to erosion and dilation by set D yields

$$(f \ominus D)(x) = \min \{f(z) : z \in D + x\} \tag{5.42}$$

$$(f \oplus D)(x) = \max \{f(z) : z \in D + x\}. \tag{5.43}$$

Thus, $f \ominus D$ is a moving-minimum over the window D and $f \oplus D$ is a moving-maximum over D.

The binary nature of erosion and dilation of signals by sets can be seen in another way. Consider a gray-scale signal f and for each possible gray level y define the **threshold set** $A[y]$ to be the set of spatial points z for which $f(z) \geq y$ (Fig. 5.12). It is immediate that $y_1 < y_2$ implies $A[y_1] \supset A[y_2]$. To erode f by a set D we desire at each x the minimum function value over the translate $D + x$, this value being $(f \ominus D)(x)$. By the definition of a moving minimum, this value is greater than or equal to y if and only if $f(z) \geq y$ for any z in $D + x$. Hence, $(f \ominus D)(x) \geq y$ if and only if $D + x$ is a subset of $A[y]$, which means $(f \ominus D)(x) \geq y$ if and only if x lies in the binary erosion $A[y] \ominus D$. Consequently, the set erosion $f \ominus D$ can be found by means of the binary erosions $A[y] \ominus D$:

$$(f \ominus D)(x) = \max \{y : x \in A[y] \ominus D\}. \tag{5.44}$$

The method of Eq. 5.44 is called **threshold decomposition**.

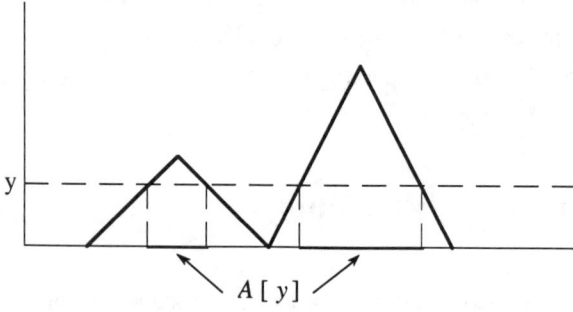

Fig. 5.12. The threshold set $A[y]$.

In fact, Eq. 5.44 is simply a restatement of the classical umbra formulation of Eq. 5.39 for flat structuring elements. To see this, suppose g is flat with support D. Then the umbra $U[g]$ is a column and a pair (x, y) lies in the binary erosion $U[f] \ominus U[g]$ if and only if the translate $D + (x, y)$ fits inside $U[f]$, which in turn means that $x \in A[y] \ominus D$. Since S in Eq. 5.39 means to take the maximum over all such y, Eqs. 5.39 and 5.44 are identical (for flat structuring elements).

5.8. Erosion Representation for Increasing Nonlinear Filters

The binary Matheron representation possesses an extension to gray-scale nonlinear operators, that is, for increasing, translation-invariant operators [63,120,121]. For any such operator Ψ, the **kernel** of Ψ is defined by

$$\text{Ker}\,[\Psi] = \{g : \Psi(g)(0) \geq 0\}. \tag{5.45}$$

A signal f lies in the kernel of Ψ if and only if the filtered signal is nonnegative at the origin. Ψ possesses the erosion representation

$$\Psi(f) = \bigvee \{f \ominus g : g \in \text{Ker}\,[\Psi]\}. \tag{5.46}$$

Thus, any increasing, translation-invariant operator possesses a morphological representation in terms of erosions. A good portion of the remainder of this text will consider certain types of such operators, the key subclass being stack filters, and, within the class of stack filters, order-statistic filters.

A **basis** for the kernel is a subclass $\text{Bas}\,[\Psi]$ of $\text{Ker}\,[\Psi]$ such that (1) there does not exist two signals $f, g \in \text{Bas}\,[\Psi]$ properly related by $\leq$ and (2) for any signal $f \in \text{Ker}\,[\Psi]$, there exists $g \in \text{Bas}\,[\Psi]$ such that $g \leq f$. The first requirement assures that there is no redundancy in the basis and the second assures that the basis is sufficient for representation. The genesis of the basis concept is straightforward: if $g_1 \leq g_2$, then $f \ominus g_1 \geq f \ominus g_2$ and therefore g_2 is not needed in the kernel expansion. If a basis exists, then it is unique and the kernel expansion reduces to the basis expansion

$$\Psi(f) = \bigvee \{f \ominus g : g \in \text{Bas}\,[\Psi]\}. \tag{5.47}$$

Given an arbitrary increasing, translation-invariant operator, it is not necessary that a basis exists; however, as in the binary setting, unless a filter is rather pathological, it will possess a basis. Filters applied in practice possess bases.

We shall go no further in the study, application, or optimization of gray-scale bases, except insofar as these pertain to nonlinear filters treated in the remainder of the text. Our purpose in introducing the gray-scale basis expansion is to have it serve as the general framework into which all increasing, nonlinear image processing naturally fits.

Finally, let us note that all transformations between lattices possess morphological (erosion, dilation) representations [8], in particular, transformations between spaces of gray-scale images. Representation takes a particularly straightforward form when signals take their values in a finite numerical set (as they do in practical digital processing) [54,55].

Chapter 6

Median and Order Statistics

Sometimes an image is so badly corrupted by noise that it is difficult to visually see any clear forms of the object in the image. In this kind of situation a simple morphological filter and even an alternating sequential filter does not work well in restoring the image. Usually a better result is obtained by methods that aim at directly estimating the gray value of a pixel from the pixel values within a window surrounding the pixel in question. The simplest estimate is just the average value of pixels within the window. The main problem of this (linear) filter is that if the window size is large enough to give good noise removal, then edges will be smoothed to such an extent that the image appears to be blurred or unfocused. This blurring effect can be avoided if instead of the average we use the median of the pixel values within the window. Median filters and their derivatives are widely used in image and signal processing and we shall describe their basic properties. Median filters are stack filters and thus are a specific type of morphological operator. In a sense they are as far as possible in the class of morphological operators from those operators using a single structuring element because, when computing the median, form information plays no role.

The one-dimensional median filter is implemented by sliding a window of odd length over the input signal one sample at a time. At each window position the samples within the window are sorted by magnitude and the centermost value, the median of the samples within the window, is the filter output. We denote the window-size by N and, as it is required to be an odd integer, write $N = 2k + 1$. Thus the one-dimensional **median** filter of length $N = 2k + 1$ can be expressed as

$$y(n) = \text{MED}\left[x(n - k), x(n - k + 1), \ldots, x(n), \ldots, x(n + k)\right], \qquad (6.1)$$

where $x(n)$ and $y(n)$ are the input and output signals. The operation defined by Eq. 6.1 is often called the running median.

Let us denote the samples in the window by $x_1, x_2, \ldots, x_{2k+1}$ and the sorted (in increasing order) list of the samples by $x_{(1)}, x_{(2)}, \ldots, x_{(2k+1)}$. Then $x_{(l)}$ is called the lth **order statistic** and the median is the $(k + 1)^{th}$ order statistic. We will briefly discuss the deterministic properties of the median that make it useful in signal processing. For simplicity we describe the properties for one-dimensional signals. The very same properties make median filtering also useful for images.

The properties that most clearly distinguish a median filter from a linear filter are its zero impulse response and ideal step response. Thus a median filter eliminates completely an isolated impulse and an ideal step edge passes through a median filter unchanged. These properties are illustrated in Fig. 6.1.

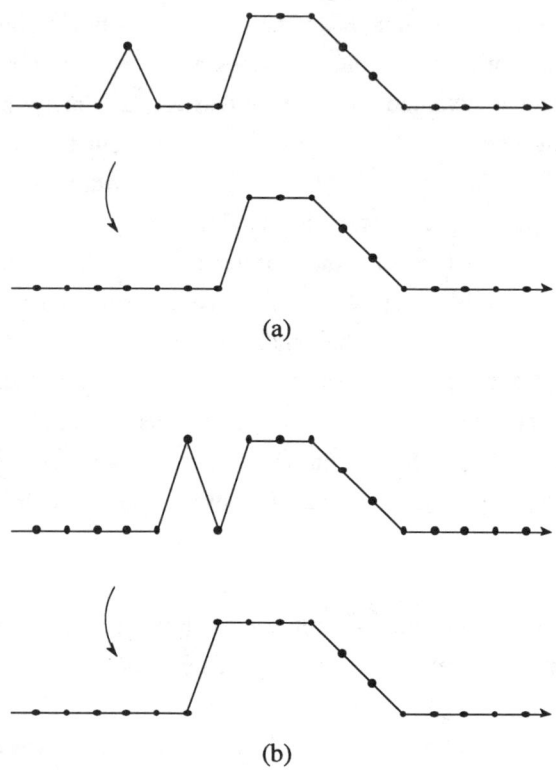

(a)

(b)

Fig. 6.1. (a) Median filter removes an isolated impulse and passes an edge unchanged. (b) An impulse close to an edge causes edge shift. The length of the filter window is 3.

The maximum duration of an impulse that is completely eliminated by a median filter depends on the filter length $N = 2k + 1$. Consider a signal $x(n)$. We call it an **impulse** of length l if

$$x(n) \begin{cases} = 0 & n < 0 \\ \neq 0 & n = 0 \\ \neq 0 & n = l - 1 \\ = 0 & n \geq l. \end{cases}$$

It is easy to see that if $l \leq k$ then the output is identically zero, i.e., the impulse is completely eliminated. On the other hand, if a signal only consists of constant sections each having length at least $k + 1$, it will not be changed by a median filter of length $2k+1$ (or less). Signals that are invariant to a median filter are called **root signals** of that median filter [60,62]. The concept of a root signal resembles the notion of a passband for a linear filter and knowledge about the root signal set of a median-type filter often provides valuable theoretical and practical information regarding the filter. Complete analysis of the root signal sets of one-dimensional median filters is given in the literature [17,59]. A simple sufficient condition for a signal to be a root of a median filter of length $2k + 1$ is that it be "locally monotonic of order $k + 1$," that is, each section of length $k + 1$ is monotonic.

The existence of nontrivial root signals is connected to the fact that the output of the median filter is always one of the input samples currently in the filter window. This property also gives rise to a drawback of median filters, which, in image processing, can be especially annoying. The median filter not only smooths the noise for homogeneous signal regions but also tends to produce regions of constant signal level, called streaks [15]. In image processing the shape of these blotches depends on the shape of the filter window. This phenomenon is clearly visible in the median filtered images of Fig. 6.4.

6.1. Two-Dimensional Median

Since the order of the samples within the filter window is irrelevant, median filtering of two-dimensional signals (images) is defined just as for one-dimensional signals. Consider an image $x(m, n)$, $1 \leq m, n \leq M$, where the indexing is according to the square matrix of Fig 6.2.

$x(1,1)$	$x(1,2)$	$\ldots$	$x(1,M)$
$x(2,1)$	$x(2,2)$	$\ldots$	$x(2,M)$
$\vdots$	$\vdots$	$\vdots$	$\vdots$
$x(M,1)$	$x(M,2)$	$\ldots$	$x(M,M)$

Fig. 6.2. The indexing of pixels.

Filtering the image by a 5-point median with cross-shaped window yields the operation

$$y(m, n) = \mathrm{MED}\,[x(m, n - 1), x(m, n), x(m, n + 1), x(m - 1, n), x(m + 1, n)]. \quad (6.2)$$

The borders of the image must be handled in some suitable way. There are several possibilities. First, we can define the output only for $2 \leq m, n \leq M - 1$. Second, we can define $x(m, n) = 0$ (or something else) outside the range $1 \leq m, n \leq M$. Third, we can extend the image periodically in both directions; i.e.,

$$x(m, n) = x(m + M, n + M) \quad \text{for all } (m, n) \in \mathbf{Z}^2. \tag{6.3}$$

The deterministic properties of two-dimensional median filters are more difficult to analyze than the corresponding properties for one-dimensional median filters. For example, there is a complete characterization of the root signals of one-dimensional median filters but very few nontrivial results for two-dimensional median filters. The large variety of window shapes also makes the formulation of general theorems difficult.

The shape of the window (or mask) has considerable effect on the two-dimensional median filtering operation. The most common masks are the vertical bar, horizontal bar, cross, X-shaped mask, square, or circle. These are depicted in Fig. 6.3. Of these basic filter windows the square is the least sensitive to image details: it filters out thin lines and cuts corners of edges. It also often produces annoying streaking. The cross filter preserves thin vertical and horizontal lines but filters out diagonal lines, whereas the X-shaped window preserves only diagonal lines. For humans, the use of the cross window is more pleasing because horizontal and vertical lines are more important in human vision. The effects of median filters with different masks are illustrated in Fig. 6.4. For sake of comparison to simple linear filtering, the results of a 5-point cross and 3×3 square moving average are also presented.

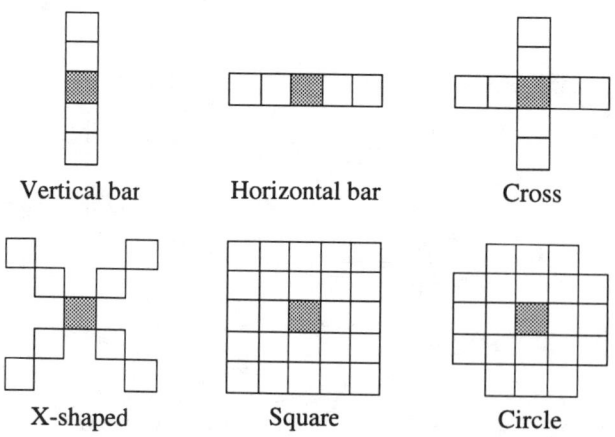

Vertical bar   Horizontal bar   Cross

X-shaped   Square   Circle

Fig. 6.3. Some common mask shapes in median filtering.

Fig. 6.4. (a) The effect of median and mean filter on a noisy image. Original image.

Fig. 6.4. (b) The effect of median and mean filter on a noisy image. Original with 10% impulsive noise added.

Fig. 6.4. (c) The effect of median and mean filter on a noisy image. Mean filtered by 5-point cross.

Fig. 6.4. (d) The effect of median and mean filter on a noisy image. Mean filtered by 9-point square.

Fig. 6.4. (e) The effect of median and mean filter on a noisy image. Median filtered by 5-point cross.

Fig. 6.4. (f) The effect of median and mean filter on a noisy image. Median filtered by 9-point square.

Fig. 6.4. (g) The effect of median and mean filter on a noisy image. Median filtered by 25-point square.

6.2. Algorithms

Median-type filters have been most useful in image and biomedical signal processing, where the amounts of data are large. This means that it is important that there are efficient algorithms for implementing these filters. There is substantial literature on the problem of computing order statistics [1,6,7,23,64,86,92,139] and we shall only describe a few simple algorithms. Although these algorithms are not asymptotically the fastest possible, their simplicity makes them perhaps best for practical purposes.

The simplest way to find the median of $N = 2k + 1$ elements is to find the smallest, delete it, again find the smallest, etc., until the median is reached. This is computationally an $O(N^2)$ method with respect to comparisons. The well-known $O(N\log N)$ sorting algorithms and sophisticated $O(N)$ algorithms for finding the median are useful only when the size of the array to be sorted is fairly large. In median filtering, even though the total amount of data may be huge, the window size is usually small.

The methods above only assume that the data values are linearly ordered. In many signal processing applications the number of possible values for the samples is relatively small; for example, the number of gray levels in image processing is typically 256. This makes methods based on a local histogram attractive. The histogram of the data in the window can be constructed with one pass over the window, and the median (or any other order statistic) is found by summing the values in the histogram until the desired order statistic is reached.

As an illustration, suppose we have the 5×5 window of data having gray-levels $0, 1, \ldots, 7$ and a histogram as depicted in Fig. 6.5.

1	2	3	3	4
0	1	1	2	1
1	0	0	1	2
7	7	6	0	1
6	6	5	5	0

0	1	2	3	4	5	6	7
5	7	3	2	1	2	3	2

Fig. 6.5. 5×5 **window of data and its histogram.**

Summing the histogram values until we reach 13, we get $5 < 13$, $5 + 7 = 12 < 13$, and $5 + 7 + 3 = 15 \geq 13$. Thus the median is 2. The complexity of this method is proportional to $N + M$, where N is the window size and M is the number of gray levels.

In one-dimensional median filtering, where the filter window slides over the signal, there is only one new input sample in the window at each time instant. This observation is utilized in running median algorithms [1,86]. The simplest running median algorithms for linearly ordered data maintain the input samples in a sorted array. As the new sample enters the window, the oldest sample is deleted from the sorted array and the new one is inserted into the proper place. This can be done using a straightforward search in $\mathbf{O}(N)$ time. Using a more complicated data structure, the asymptotic complexity drops to $\mathbf{O}(\log N)$. Because of the extra overhead required by this data structure, this method performs better in practice only for very long windows.

If the number of values that the data can assume is small, histogram-based methods are best for computing running order statistics. We need only update the histogram at each time instant. This can easily be done in constant time; that is, its complexity is independent of the window size.

For two-dimensional signals the computation of running order statistics is more time consuming because at each time instant there are (for an $N \times N$ window) N new samples in the window and the N oldest have been discarded. The computation can be done by inserting and discarding samples one by one. For histogram methods it is better to first update the histogram with all new samples and then search the new output value [86].

If the $N \times N$ window moves line by line, there are N new samples at each time instant compared to the previous window position; however, apart from the border areas of the image, only one sample is such that it has not been in *any* previous window. This fact is utilized in a clever algorithm [64] that computes the running two-dimensional median in constant time on the average.

Chapter 7

Stack Filters

In this chapter we look at some deterministic properties of the median filter and show that this leads to another very useful generalization of the median filter, namely, the stack filter [211]. The stack filter expression in turn provides us with a general method to analyze the statistical properties of weighted median filters and even some morphological filters.

Stack filters are, in fact, (increasing) morphological operators that are defined by flat structuring elements. This means that there have been two independent paths in image processing both leading to the same filter class. This is a clear indication of how natural and useful the window logic approach is in nonlinear digital image processing.

In the definition of stack filters we shall follow the route by which they were originally derived, even though we could use the machinery of gray-scale morphology. We feel this approach gives additional insight into the image processing problem and also emphasizes the close connection to order-statistics-based digital filtering techniques.

Considering the median filter as a function $\mu : \mathbf{R}^{2k+1} \to \mathbf{R}$, we see that it has the following properties: (i) for any $a \in \mathbf{R}$,

$$\mu(x_1 + a, \ldots, x_{2k+1} + a) = \mu(x_1, \ldots, x_{2k+1}) + a, \qquad (7.1)$$

(ii) for any $a \in \mathbf{R}$,

$$\mu(ax_1, \ldots, ax_{2k+1}) = a\mu(x_1, \ldots, x_{2k+1}). \qquad (7.2)$$

On the other hand, in general,

$$\mu(x_1 + y_1, \ldots, x_{2k+1} + y_{2k+1}) \neq \mu(x_1, \ldots, x_{2k+1}) + \mu(y_1, \ldots, y_{2k+1}), \qquad (7.3)$$

and thus the mapping μ is, of course, not linear. Property (i) says that the median operation is spatially translation invariant and property (ii) says that it is invariant to scaling by any real number. In fact, a far stronger invariance holds. Let $\xi : \mathbf{R} \to \mathbf{R}$ be monotonic. Then

$$\mu(\xi(x_1), \ldots, \xi(x_{2k+1})) = \xi(\mu(x_1, \ldots, x_{2k+1})). \qquad (7.4)$$

Let ξ_a be defined by

$$\xi_a(t) = \begin{cases} 1 & \text{if } t > a \\ 0 & \text{otherwise.} \end{cases} \tag{7.5}$$

Applying Eq. 7.4 with ξ_a we find

$$\mu(\xi_a(x_1), \ldots, \xi_a(x_{2k+1})) = \xi_a(\mu(x_1, \ldots, x_{2k+1})). \tag{7.6}$$

Equation 7.6 possesses the following interpretation: if we want to know whether $\mu(x_1, \ldots, x_{2k+1})$ is greater than a, it is enough to check if at least $k + 1$ of the x_i are greater than a. The fact that Eq. 7.6 can be understood as a Boolean function leads to an important generalization of the median operation, namely, the stack filter.

It is generally very difficult to analyze or optimally design nonlinear filters. Almost all the powerful tools and elegant methods of linear digital signal processing are completely useless when dealing with nonlinear methods. However, the output of the median filter is decided solely on the basis of the ranks of the samples. This property makes it possible to use a powerful technique, called threshold decomposition, to divide the analysis into smaller and simpler parts. Using threshold decomposition we can, via Eq. 7.4, derive all properties of median filters by just studying their effect on binary signals. Threshold decomposition and its relation to mathematical morphology was discussed in Section 5.7.

Threshold decomposition of an M-valued signal $x(n)$, where the samples are integer valued and $0 \leq x(n) \leq M - 1$, means decomposing it into $M - 1$ binary signals $x^{(1)}(n), \ldots x^{(M-1)}(n)$ according to the thresholding rule

$$x^{(m)}(n) = \begin{cases} 1 & \text{if } x(n) \geq m \\ 0 & \text{otherwise.} \end{cases} \tag{7.7}$$

Let $\mathbf{u}$ and $\mathbf{v}$ be binary signals (sequences) of fixed length. Define

$$\mathbf{u} \leq \mathbf{v} \quad \text{iff} \quad u(n) \leq v(n) \quad \text{for all } n. \tag{7.8}$$

As the relation defined by Eq. 7.8 is reflexive, antisymmetric, and transitive, it defines a partial order in the set of binary signals of fixed length. Now, consider a signal $\mathbf{x}$ and its thresholded binary signals $\mathbf{x}^{(1)}, \ldots, \mathbf{x}^{(M-1)}$. Clearly,

$$\mathbf{x}^{(i)} \leq \mathbf{x}^{(j)} \quad \text{if} \quad i \geq j. \tag{7.9}$$

Thus the binary signals $\mathbf{x}^{(1)}, \ldots, \mathbf{x}^{(M-1)}$ are nonincreasing in the sense of the partial ordering of Eq. 7.8.

As an example, consider a five-level ($M = 5$) integer-valued signal $x(n)$, (that is, x is a mapping $x : \mathbf{Z} \to \{0, 1, 2, 3, 4\}$), and its section $0 \le n \le 10$:

$$x(n) = \quad \ldots 0\ 0\ 1\ 3\ 2\ 4\ 4\ 0\ 0\ 4\ 0 \ldots . \tag{7.10}$$

Its thresholded binary signals are given by

$$x^{(4)}(n) = \quad \ldots 0\ 0\ 0\ 0\ 0\ 1\ 1\ 0\ 0\ 1\ 0 \ldots$$

$$x^{(3)}(n) = \quad \ldots 0\ 0\ 0\ 1\ 0\ 1\ 1\ 0\ 0\ 1\ 0 \ldots$$

$$x^{(2)}(n) = \quad \ldots 0\ 0\ 0\ 1\ 1\ 1\ 1\ 0\ 0\ 1\ 0 \ldots \tag{7.11}$$

$$x^{(1)}(n) = \quad \ldots 0\ 0\ 1\ 1\ 1\ 1\ 1\ 0\ 0\ 1\ 0 \ldots .$$

From Eq. 7.11, and more generally from Eq. 7.7, we see that

$$x(n) = \sum_{i=0}^{M-1} x^{(i)}(n). \tag{7.12}$$

This illustrates the fact, which follows from Eq. 7.4, that the median filtering of Eq. 7.10 can be done by separately filtering the thresholded signals and then adding them. We can write this as the following three-stage procedure:

1) Decompose the signal into $M - 1$ binary signals according to Eq. 7.7.

2) Apply the filter to the binary signals separately.

3) Add the filtered signals.

This procedure is depicted in Fig. 7.1.

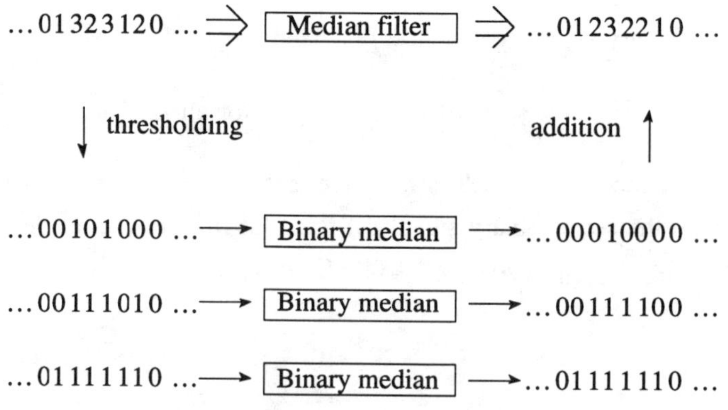

$\ldots 0\,1\,3\,2\,3\,1\,2\,0 \ldots \Rightarrow \boxed{\text{Median filter}} \Rightarrow \ldots 0\,1\,2\,3\,2\,2\,1\,0 \ldots$

$\downarrow$ thresholding   addition $\uparrow$

$\ldots 0\,0\,1\,0\,1\,0\,0\,0 \ldots \longrightarrow \boxed{\text{Binary median}} \longrightarrow \ldots 0\,0\,0\,1\,0\,0\,0\,0 \ldots$

$\ldots 0\,0\,1\,1\,1\,0\,1\,0 \ldots \longrightarrow \boxed{\text{Binary median}} \longrightarrow \ldots 0\,0\,1\,1\,1\,1\,0\,0 \ldots$

$\ldots 0\,1\,1\,1\,1\,1\,1\,0 \ldots \longrightarrow \boxed{\text{Binary median}} \longrightarrow \ldots 0\,1\,1\,1\,1\,1\,1\,0 \ldots$

Fig. 7.1. Filtering by thresholding.

From threshold decomposition we know that the thresholded binary signals satisfy

$$\mathbf{x}^{(i)} \le \mathbf{x}^{(j)} \ \ if \ \ i \ge j.$$

Property 7.6 implies that also the binary output signals $\mathbf{y}^{(i)}$ satisfy

$$\mathbf{y}^{(i)} \le \mathbf{y}^{(j)} \ \ if \ \ i \ge j. \tag{7.13}$$

Median filtering of a binary signal is essentially a Boolean function of the $2k+1$ variables inside the filter window. For example, the binary median filtering with window length three can be written as

$$y(n) = x(n-1)x(n) + x(n-1)x(n+1) + x(n)x(n+1),$$

where $+$ and $\cdot$ denote Boolean OR and AND operations.

At this point a natural question arises: Can we use any Boolean function to define a filtering operation via threshold decomposition? It turns out that to obtain useful properties the class of Boolean functions must be restricted in the way that the binary output signals $\mathbf{y}^{(i)}$ satisfy Eq. 7.13, which essentially means that the "stack" of output binary signals corresponds to the threshold decomposition of some M valued signal. This suitably restricted class of Boolean functions, namely, positive Boolean functions, form the basis of stack filters first introduced in Ref. [211].

Let f be a Boolean function with arguments indexed from $-k$ to k, where k is a nonnegative integer, and use the notation $\mathbf{x} = (x_{-k}, \ldots, x_k)$ for the elements of $\{0,1\}^{2k+1}$, the set of all $\{0,1\}$ vectors of length $2k+1$. Now, f is called a **positive Boolean function** if it satisfies

$$\mathbf{u} \le \mathbf{v} \Rightarrow f(\mathbf{u}) \le f(\mathbf{v}).$$

It is well known, cf. [106,136], that a Boolean function is positive if and only if it contains no complemented variables in its minimum sum of products form.

As an example, let $k = 1$. The functions

$$f(\mathbf{x}) = x_{-1}x_0 + x_{-1}x_1 + x_0x_1$$

and

$$f(\mathbf{x}) = x_0 + x_{-1}x_1$$

are positive Boolean functions.

A M-valued stack filter is defined using threshold decomposition and a positive Boolean function $f(\mathbf{x}) = f(x_{-k}, \ldots, x_k)$ as follows. Let $\mathbf{x}$ be an M-valued input signal, that is, $x(n) \in \{0, 1, \ldots, M-1\}$ for all $n \in \mathbf{Z}$. Form the thresholded binary signals $\mathbf{x}^{(1)}, \ldots, \mathbf{x}^{(M-1)}$ by

$$x^{(i)}(n) = \begin{cases} 1 & \text{if } x(n) \geq i \\ 0 & \text{otherwise.} \end{cases} \tag{7.14}$$

The output of filtering the ith thresholded signal $\mathbf{x}^{(i)}$ at point n is defined by

$$y^{(i)}(n) = f(x^{(i)}(n-k), \ldots, x^{(i)}(n+k)), \tag{7.15}$$

where $x^{(i)}(n+j)$, $-k \leq j \leq k$, are understood as Boolean variables. The output of the stack filter defined by f at point n is now

$$y(n) = \sum_{i=1}^{M-1} y^{(i)}(n), \tag{7.16}$$

where the values $y^{(i)}(n)$ of Boolean functions are now understood as real 0's and 1's. This is illustrated in Fig. 7.2

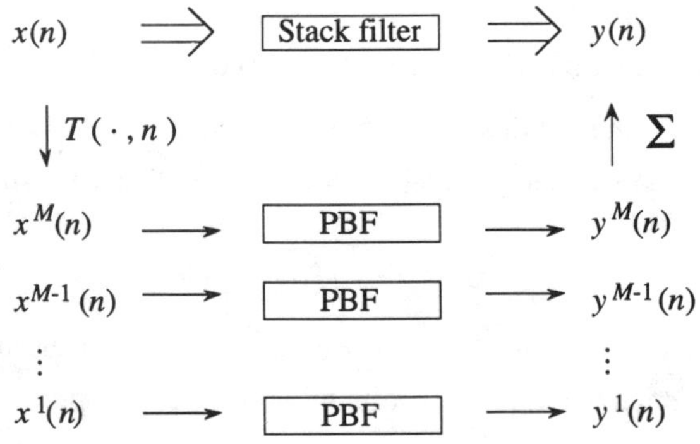

$$x(n) \quad \Longrightarrow \quad \boxed{\text{Stack filter}} \quad \Longrightarrow \quad y(n)$$

$$\downarrow T(\cdot, n) \qquad\qquad\qquad\qquad \uparrow \Sigma$$

$$x^M(n) \quad \longrightarrow \quad \boxed{\text{PBF}} \quad \longrightarrow \quad y^M(n)$$

$$x^{M\text{-}1}(n) \quad \longrightarrow \quad \boxed{\text{PBF}} \quad \longrightarrow \quad y^{M\text{-}1}(n)$$

$$\vdots \qquad\qquad\qquad\qquad\qquad \vdots$$

$$x^1(n) \quad \longrightarrow \quad \boxed{\text{PBF}} \quad \longrightarrow \quad y^1(n)$$

Fig. 7.2. Block diagram of stack filter.

To illustrate the method, consider the Boolean function $f(\mathbf{x}) = x_{-1}x_1 + x_0$ and a 5-valued signal segment $\ldots 0\ 0\ 2\ 1\ 2\ 3\ 4\ 0\ 1\ 0\ 0\ldots$. The stack filtering can be expressed schematically as in Fig. 7.3. Notice that in gray-scale values f corresponds to the expression $\max\left(\min\left(x(n-1), x(n+1)\right), x(n)\right)$.

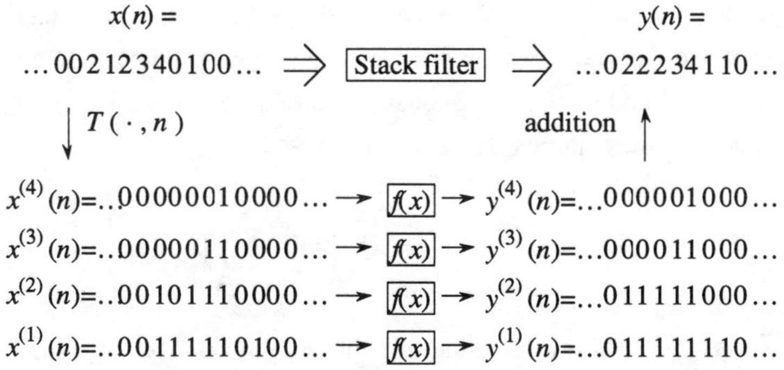

$$x(n) = \qquad\qquad\qquad\qquad\qquad y(n) =$$

$$\ldots 0\,0\,2\,1\,2\,3\,4\,0\,1\,0\,0\ldots \;\Longrightarrow\; \boxed{\text{Stack filter}} \;\Longrightarrow\; \ldots 0\,2\,2\,2\,3\,4\,1\,1\,0\ldots$$

$$\Big\downarrow T(\,\cdot\,,n) \qquad\qquad\qquad\qquad\qquad \text{addition} \Big\uparrow$$

$$x^{(4)}(n)=..\,0\,0\,0\,0\,0\,0\,1\,0\,0\,0\,0\ldots \rightarrow \boxed{f(x)} \rightarrow y^{(4)}(n)=\ldots 0\,0\,0\,0\,0\,1\,0\,0\,0\ldots$$

$$x^{(3)}(n)=..\,0\,0\,0\,0\,0\,1\,1\,0\,0\,0\,0\ldots \rightarrow \boxed{f(x)} \rightarrow y^{(3)}(n)=\ldots 0\,0\,0\,0\,1\,1\,0\,0\,0\ldots$$

$$x^{(2)}(n)=..\,0\,0\,1\,0\,1\,1\,1\,0\,0\,0\,0\ldots \rightarrow \boxed{f(x)} \rightarrow y^{(2)}(n)=\ldots 0\,1\,1\,1\,1\,1\,0\,0\,0\ldots$$

$$x^{(1)}(n)=..\,0\,0\,1\,1\,1\,1\,1\,0\,1\,0\,0\ldots \rightarrow \boxed{f(x)} \rightarrow y^{(1)}(n)=\ldots 0\,1\,1\,1\,1\,1\,1\,1\,0\ldots$$

Fig. 7.3. Illustration of stack filtering operation using threshold decomposition.

An important class of stack filters is defined via a generalization of the median filter. With weighted median filters, we assign different emphasis to different samples in the window by duplicating them suitably many times. As an example assume that the elements in the window are x_{-2}, x_{-1}, x_0, x_1, and x_2. We can put more weight to the samples on and close to the center of the window by repeating x_0 three (say) times and x_{-1} and x_1 two times and then take the median over the expanded set $x_{-2}, x_{-1}, x_{-1}, x_0, x_0, x_0, x_1, x_1, x_2$ of nine elements. If the sample x_i is duplicated w_i times, this is compactly expressed as

$$\text{WM}\left[w_{-k} \diamond x_{-k}, w_{-k+1} \diamond x_{-k+1}, \ldots, w_k \diamond x_k\right]$$

and is called the **weighted median** of $x_{-k}, \ldots, x_k$ with weights $w_{-k}, \ldots, w_k$.

Weighted median filters often balance between detail preservation and noise reduction. As an example consider the weight masks shown in Fig. 7.4. Giving more weight to the center increases detail preservation and by choosing the weights suitably, for instance, vertical and horizontal lines can be preserved.

Mask

1	1	1
1	3	1
1	1	1

1	3	1
3	5	3
1	3	1

Details preserved   Horizontal/vertical corners   Horizontal/vertical corners

45° tilted corners   Horizontal/vertical lines

(a)   (b)

Fig. 7.4. Masks for weighted median filters. The center weighted median (a) preserves both horizontal/vertical corners and 45° tilted corners but not any lines of width one pixel. The weighted median with mask (b) does not preserve 45° tilted corners but preserves horizontal/vertical lines of width one pixel.

In the following we shall see that the weighted median filter is, in fact, a stack filter where the positive Boolean function is of a special type, namely, self dual and linearly separable. It is well known that the median of an odd number of numbers minimizes the sum of distances to these numbers; i.e.

$$\text{MED}\,[x_{-k},\ldots,x_k] = \arg\left\{\min_t \sum_{i=-k}^{k} |x_i - t|\right\}, \tag{7.17}$$

where $\arg\{\min h(t)\}$ denotes the value of t that minimizes $h(t)$. Consider the computation of the weighted median of length $2k+1$ with weights $w_{-k},\ldots,w_k$. Because the weighting means duplication we only need to take the term $|t - x_i|$ in Eq. 7.17 w_i times to obtain a similar expression for the weighted median filter. Thus

$$\text{WM}\,[w_{-k} \diamond x_{-k},\ldots,w_k \diamond x_k] = \arg\left\{\min_t \sum_{i=-k}^{k} w_i|x_i - t|\right\}.$$

The objective function

$$s(t) = \sum_{-k}^{k} w_i|x_i - t|$$

is everywhere continuous, and differentiable if $t \notin \{x_{-k},\ldots,x_k\}$. Its derivative is

$$s'(t) = -\sum_{(x_i < t)} w_i + \sum_{(x_i > t)} w_i. \tag{7.18}$$

From Eq. 7.18 we see that $s(t)$ has a unique minimum for any choice of $\{x_{-k},\ldots,x_k\}$ if and only if there is no partition

$$\{x_{-k},\ldots,x_k\} = A \cup B, \quad A \cap B = \emptyset,$$

such that

$$\sum_{i \in A} w_i = \sum_{i \in B} w_i. \tag{7.19}$$

Writing Eq. 7.19 as

$$2\sum_{i \in A} w_i = \sum_{i \in A} w_i + \sum_{i \in B} w_i = \sum_{i=-k}^{k} w_i \tag{7.20}$$

we see that a sufficient condition for a weighted median with positive integer weights to be unique is that the sum of weights is odd, since, were Eq. 7.20 to hold, it would mean that the sum of the weights was even.

Equation 7.18 gives a straightforward method to compute WM $[w_{-k} \diamond x_{-k}, \ldots, w_k \diamond x_k]$: Denote by $[(-k), (-k+1), \ldots, (k)]$ a permutation of $(-k, -k+1, \ldots, k)$ that arranges $x_{-k}, \ldots, x_k$ into an increasing order; that is

$$x_{(-k)} \leq x_{(-k+1)} \leq \cdots \leq x_{(k)}.$$

Compute the sums

$$\sum_{i=-k}^{l} w_{(i)}, \quad l = -k, -k+1, \ldots \tag{7.21}$$

and let l_0 be the smallest index such that

$$\sum_{i=-k}^{l} w_{(i)} \geq \frac{1}{2} \sum_{i=-k}^{k} w_{(i)}. \tag{7.22}$$

Then $x_{l_0} = \text{WM} [w_{-k} \diamond x_{-k}, \ldots, w_k \diamond x_k]$.

As an example, compute WM $[1 \diamond 3, 2 \diamond 2, 3 \diamond 1, 2 \diamond 5, 1 \diamond 4]$. The permutation that orders the elements is $[(-2), (-1), (0), (1), (2)] = (0, -1, -2, 2, 1)$ and

$$\sum_{i=-2}^{-2} w_{(i)} = 3, \quad \sum_{i=-2}^{-1} w_{(i)} = 5 \geq \frac{9}{2}. \tag{7.23}$$

Thus WM $[1 \diamond 3, 2 \diamond 2, 3 \diamond 1, 2 \diamond 5, 1 \diamond 4] = x_{(-1)} = x_{-1} = 2$.

Consider the Boolean function $f(\mathbf{x})$ that will realize WM $[w_{-k} \diamond x_{-k}, \ldots, w_k \diamond x_k]$ on each level of threshold decomposition. From Eqs. 7.20 and 7.21 we see that it can be written in the form

$$f(x_{-k}, \ldots, x_k) = \begin{cases} 1 & \text{if } \sum_{i=-k}^{k} w_i x_i > \frac{1}{2} \sum_{i=-k}^{k} w_i. \\ 0 & \text{otherwise.} \end{cases} \tag{7.24}$$

A Boolean function that can be expressed in the form

$$f(x_{-k}, \ldots, x_k) = \begin{cases} 1 & \text{if } \sum_{i=-k}^{k} w_i x_i > T \\ 0 & \text{otherwise} \end{cases} \tag{7.25}$$

is called **linearly separable**. A Boolean function $f(\mathbf{x}) = f(x_{-k}, \ldots, x_k)$ satisfying

$$\overline{f}(\overline{\mathbf{x}}) = f(\mathbf{x}) \quad \text{for all } \mathbf{x}, \tag{7.26}$$

where $\overline{\mathbf{x}}$ means the negation of $\mathbf{x}$, is called **self dual**. By direct computation we see that any Boolean function of the form of Eq. 7.24, where w_i are positive integers with $\sum w_i$ odd, is self dual: Let $\mathbf{x} \in \{0, 1\}^{2k+1}$. The following chain of equivalences holds:

$$f(\mathbf{x}) = 1 \Leftrightarrow \sum_{i=-k}^{k} w_i x_i > \frac{1}{2} \sum_{i=-k}^{k} w_i$$

$$\Leftrightarrow \sum_{i=-k}^{k} w_i(1 - x_i) < \frac{1}{2} \sum_{i=-k}^{k} w_i \qquad (7.27)$$

$$\Leftrightarrow f(\overline{\mathbf{x}}) = 0,$$

proving the assertion.

There are two very important advantages of the stack filter expression of median-type filters. First, the analysis of iterated or cascaded operations becomes, in a sense, trivial because we need only compute the Boolean function obtained by substituting a shifted version of the function into each variable. Second, the statistical analysis of median and stack filters can be treated in a unified way.

For example, consider the operation resulting when a three-point median is applied twice to a signal $x(n)$. The corresponding Boolean function is $f(x_{-1}, x_0, x_1) = x_{-1}x_0 + x_{-1}x_1 + x_0x_1$. Thus, after one pass of the three-point median the resulting signal can be expressed (with Boolean operations) as

$$y(n) = x(n-1)x(n) + x(n-1)x(n+1) + x(n)x(n+1) \qquad (7.28)$$

and thus the final result is

$$z(n) = y(n-1)y(n) + y(n-1)y(n+1) + y(n)y(n+1)$$

$$= x(n-2)x(n-1)x(n+1) + x(n-2)x(n)x(n+2) + x(n-1)x(n)$$

$$+ x(n-1)x(n+1)x(n-2) + x(n)x(n+1).$$

It is interesting that this particular Boolean function is linearly separable and the resulting stack filter is a weighted median,

$$z(n) = \text{WM}\,[1 \diamond x(n-2), 2 \diamond x(n-1), 3 \diamond x(n), 2 \diamond x(n+1), 1 \diamond x(n+2)]. \qquad (7.29)$$

For another example, consider the **center-weighted median** filter

$$y(n) = \text{WM}\left[1 \diamond x(n-k), \ldots 1 \diamond x(n-1), (2k-1) \diamond x(n), 1 \diamond x(n+1), \ldots 1 \diamond x(n+k)\right].$$
$$(7.30)$$

The corresponding Boolean function can be written as

$$f(x_{-k}, \ldots, x_k) = x_0 \sum_{i=1}^{k}(x_{-i} + x_i) + \prod_{i=1}^{k}(x_{-i}x_i). \qquad (7.31)$$

Straightforward calculation shows that if appropriately shifted versions of this Boolean function are substituted into the variables, then the resulting function collapses back to its original form in Eq. 7.29. This shows that the center-weighted median is an idempotent operation. The result holds also for two- and higher-dimensional center-weighted median filters of the above type [70]. A more detailed exposition of cascading weighted median and stack filters can be found in Ref. [221].

Chapter 8

Statistical Properties of Median-type and Stack Filters

Median-type filters were originally introduced as data smoothers for situations where traditional linear smoothers do not perform well, for example, if the noise is impulsive or if there are abrupt changes in the level or amplitude of the signal. The performance of the filter depends on how well it can suppress the undesired part of the signal while retaining the desired part. Because median-type filters are nonlinear, it is very difficult or perhaps impossible to derive general results that would accurately describe the statistical behavior for a wide range of random signals as is the case for linear filters. Examples of characterizations that are useful in practice and possible to compute are white-noise attenuation and response to a noisy step or edge signal. Because the effects of a nonlinear filter to the noise and to the signal cannot be separated as with linear filters, an apparently easy task, such as the analysis of filter response to a noisy edge, may be almost impossible. However, theoretical work and extensive simulations have provided a fairly good understanding of the statistical behavior of median-type filters.

8.1. Noise Attenuation

The analysis of the effects of a median filter on pure white noise is fairly simple because the distribution function of the median of independent and identically distributed random variables is easily derived. Let $X_1, \ldots, X_{2k+1}$ be independent and identically distributed random variables with a common cumulative distribution function $F(t)$ and density $f(t)$. Consider the event

$$X_{i_1}, \ldots, X_{i_k} \leq t, \quad t \leq X_{i_{k+1}} \leq t + dt, \quad t < X_{i_{k+2}}, \ldots, X_{i_{2k+1}}, \qquad (8.1)$$

where $i_j \neq i_l$ for $j \neq l$. The probability of this event is

$$F(t)^k (1 - F(t))^k f(t)\, dt. \qquad (8.2)$$

As there are $\frac{(2k+1)!}{k!k!}$ disjoint events of the form given in Eq. 8.1 that together form the event

$$t \leq \text{MED}\,[X_1, \ldots, X_{2k+1}] \leq t + dt,$$

we see that the density function $g(t)$ of the median is

$$g(t) = \frac{(2k+1)!}{k!k!} f(t) F(t)^k (1 - F(t))^k. \qquad (8.3)$$

Integrating, we get the distribution function

$$G(t) = \sum_{i=k+1}^{2k+1} \binom{2k+1}{i} F(t)^i (1 - F(t))^{2k+1-i}. \tag{8.4}$$

From Eqs. 8.3 and 8.4 all statistical properties of median filters on white noise can be computed either in closed form or by numerical methods. It can be shown that under some very general assumptions [cf. 93] the asymptotic distribution ($N \to \infty$) of the median is normal with expectation

$$\mu_{med} = x_{0.5},$$

where $x_{0.5}$ is defined by $F(x_{0.5}) = \frac{1}{2}$, and

$$\sigma^2_{med} = \frac{1}{4Nf(x_{0.5})^2}. \tag{8.5}$$

The important observation that follows from Eq. 8.5 is that the asymptotic ratio of the variances of average and median is

$$\frac{\sigma^2_{ave}}{\sigma^2_{med}} = 4\sigma^2 f(x_{0.5})^2,$$

where σ^2 is the variance of the noise. It shows that even for white Gaussian noise, for which the mean is the optimal smoother, the variance of the median is only 57% larger than the variance of the mean. The asymptotic variances of the mean and the median are summarized in Table 8.1.

Noise density	Mean	Median
Uniform $f(t) = \begin{cases} \dfrac{1}{\sqrt{12}\,\sigma}, & -\sqrt{3}\sigma \le t \le \sqrt{3}\sigma \\ 0 & \text{otherwise} \end{cases}$	$\dfrac{\sigma^2}{N}$	$\dfrac{3\sigma^2}{N+2}$
Gaussian $f(t) = \dfrac{1}{\sqrt{2\pi}\,\sigma} e^{-\frac{t^2}{2\sigma^2}}$	$\dfrac{\sigma^2}{N}$	$\dfrac{\pi\sigma^2}{2N}$
Laplacian $f(t) = \dfrac{1}{\sqrt{2}\sigma} e^{-\frac{\sqrt{2}}{\sigma}|t|}$	$\dfrac{\sigma^2}{N}$	$\dfrac{\sigma^2}{2N}$

Table 8.1. Asymptotic variances of mean and median.

An attractive property of the median filter is that it will leave a perfect edge unchanged while any linear filter will round or blur the edge. However, this is true only for noiseless edges. For noisy edges the behavior of the median resembles that of the moving average, as illustrated in Fig. 8.1.

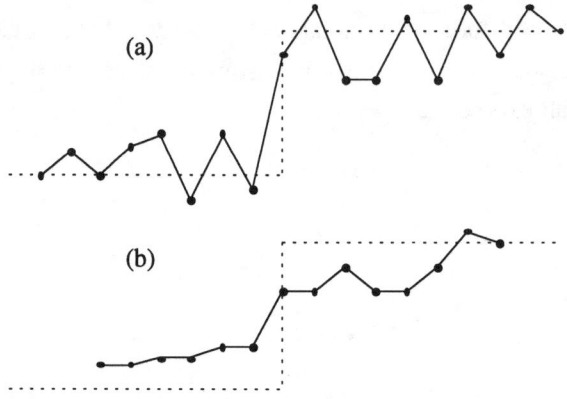

Fig. 8.1. (a) Original noisy signal, (b) median filtered signal. The window length is 5.

8.2. Median as Maximum Likelihood Estimator

It is useful to look at the median filter from the estimation theory point of view. The analysis gives insight into the median filtering operation and also leads to useful generalizations.

Many signal processing tasks can be formulated in the following way. Suppose we have the observations

$$x_i = \theta s_i + n_i, \quad i = 1, \ldots, N, \tag{8.6}$$

where $(s_1, \ldots, s_N)$ is a known signal waveform, θ is a unknown "amplitude" parameter to be estimated, and $(n_1, \ldots, n_N)$ is a sequence of independent and identically distributed (*i.i.d.*) random variables with a common distribution function $F(t)$.

Assume first that $N = 2k + 1$, $s_1 = \ldots = s_N = 1$, and that n_i, $i = 1, \ldots, N$, has a Laplace (or biexponential) distribution with probability density function

$$f(t) = \frac{\alpha}{2} e->-\alpha|t|}, \quad \alpha > 0.$$

This means that the observations form a simple random sample from a population having density

$$f(t) = \frac{\alpha}{2} e^{-\alpha|t-\theta|}, \quad \alpha > 0. \tag{8.7}$$

The maximum likelihood (ML) estimation principle says that given a simple random sample $x_1, \ldots, x_N$ we should choose as the estimate for θ the value $\hat{\theta}$ for which the joint density is maximized for this particular sample. The joint density is, by independence,

$$f(t_1, \ldots, t_N) = \prod_{i=1}^{N} f(t_i) = \left(\frac{\alpha}{2}\right)^N e^{-\alpha \sum_{i=1}^{N} |t_i - \theta|}. \tag{8.8}$$

Substituting the values x_i for t_i we see that maximizing Eq. 8.8 at $t_i = x_i$, $i = 1, \ldots, N$, is equivalent to minimizing

$$L(\theta) = \sum_{i=1}^{N} |x_i - \theta|. \tag{8.9}$$

Considering the graph of $L(\theta)$ we see that the value of θ minimizing $L(\theta)$, for which we use the shorthand notation

$$\arg \{\min_{\theta} \sum_{i=1}^{N} |x_i - \theta|\}, \tag{8.10}$$

is exactly the median of $x_1, \ldots, x_N$; that is, $\hat{\theta} = \text{MED}[x_1, \ldots, x_N]$. This means that using the median filter is equivalent to finding the ML estimate of the amplitude of a constant signal under the assumption that noise is $i.i.d.$ Laplace distributed. It is interesting that only changing the noise to $i.i.d.$ Gaussian changes Eq. 8.9 to

$$L(\theta) = \sum_{i=1}^{N} (x_i - \theta)^2 \tag{8.11}$$

which leads to the simplest linear smoother, namely, the moving average of length N,

$$\hat{\theta} = \frac{1}{N} \sum_{i=1}^{N} x_i.$$

This estimator, written as a digital filter, is

$$y(n) = \frac{1}{2k+1} \sum_{i=-k}^{k} x(n+i).$$

The interpretation of the median filter as a device producing the ML estimate for location parameter under Laplacian noise partly explains the good behavior of the median filter when impulsive noise is present. To wit, the Laplace distribution is often used to model impulsive or heavy tailed noise. As an estimator the median belongs to the class of so-called robust estimators [cf. 87], which have the property of not being sensitive to variations in the distributions of the underlying population. A simple example of the different behavior of the median and mean of the same window size is the following. If we let one sample value become arbitrarily large, then the mean will also become arbitrarily large. The median either remains the same (if the particular sample value is originally at least the median) or just moves to the next larger sample value. Also, if we compare the variances of the sample mean and sample median for large sample sizes, they depend on the distribution of the underlying population in radically different ways. For instance, changes in the variance of the underlying distribution need not affect the variance of the sample median at all. A good survey of robust methods in linear and nonlinear signal processing is given in Ref. [94].

The preceding ML approach leads to useful generalizations of the median filter in three ways. First, we can assume that the corrupting noise has density of the form

$$f(t) = \alpha e^{-\beta|t|^\gamma}, \tag{8.12}$$

where β and γ are positive constants and α is the necessary normalizing factor. This assumption on the noise distribution leads to the filtering operation [4]

$$y(n) = \arg\left\{\min_x \sum_{i=-k}^{k} |x(n+i) - x|^\gamma\right\} \tag{8.13}$$

which has several interesting properties. If $\gamma = 1$, it results in the median filter. If $\gamma = 2$, it results in the linear simple moving average. If $\gamma \to \infty$, it will approach the midrange detector, that is,

$$y(n) = \frac{1}{2}(\min\{x(n+i) : -k \le i \le k\} + \max\{x(n+i) : -k \le i \le k\}).$$

If $\gamma \le 1$, the filter will behave similarly to the median filter in the sense that its impulse response is zero and its step response is ideal. If $\gamma < 1$, the filter has an edge-enhancing property such that if used on a gray-scale image, it will increase its contrast. The general appearance of the objective function in Eq. 8.13 is plotted in Fig. 8.2 for $\gamma = 0.25, 1, 2$, and 3. Figure 8.2. also shows the edge-enhancing ($\gamma < 1$) and edge-smoothing ($\gamma > 1$) properties of the filter.

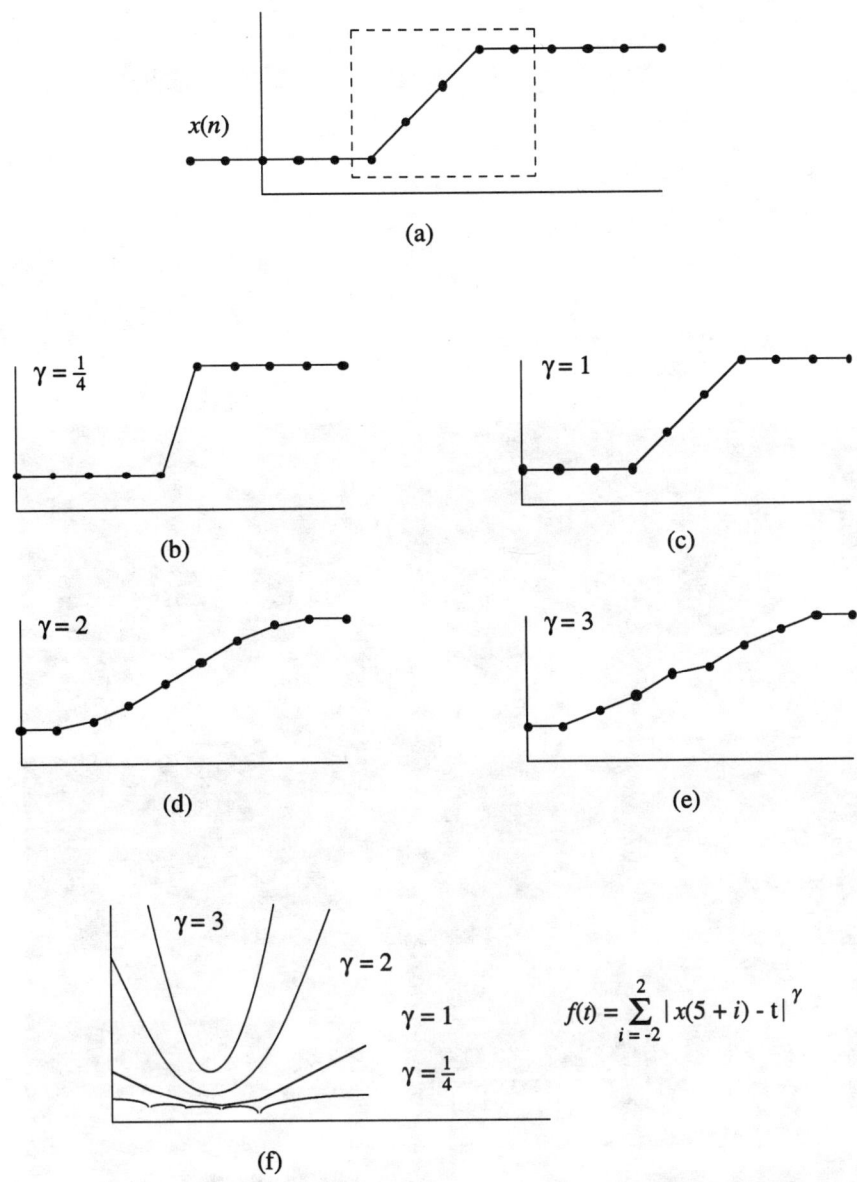

Fig. 8.2. Illustration of the edge-enhancing property of the filter defined by Eq. 8.13 with $n = 5$: (a) original edge, (b) filtered with $\gamma = \frac{1}{4}$, (c) filtered with $\gamma = 1$, (d) filtered with $\gamma = 2$, (e) filtered with $\gamma = 3$, (f) form of the objective function of Eq. 8.13 for various values of γ.

When $\gamma < 1$, the filter has the tendency of producing large areas of constant gray-level that can be utilized, for example, in image coding. This effect is shown in Fig. 8.3.

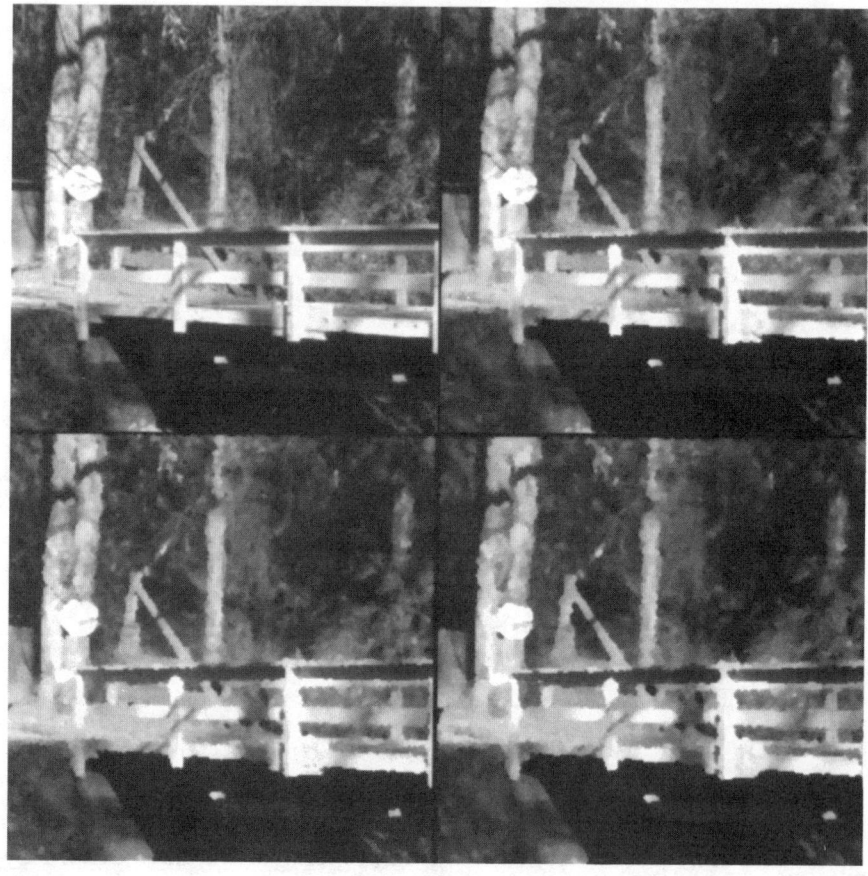

Fig. 8.3. The behavior of the filter defined by Eq. 8.13 for $\gamma = 0.1$. First row: original and the result after one filtering. Second row: the result after two filterings and after five filterings.

A second generalization is obtained in the following manner. Again assume in Eq. 8.6 that $s_1 = \ldots = s_N = 1$ and the independence of noise components, but allow them to have centered Laplace densities with different variances. Let the noise component n_i come from a population having density

$$f_i(t) = \frac{\gamma_i}{2} e^{-\gamma_i |t|}, \; i = 1, \ldots, N.$$

This implies that the ML estimate for θ is

$$\hat{\theta} = \arg \{ \min_{\theta} \sum_{i=1}^{N} \gamma_i |x_i - \theta| \}. \tag{8.14}$$

To clarify the meaning of Eq. 8.14 assume that the γ_i are positive integers. Comparing to Eq. 8.10 we see that $\hat{\theta}$ in Eq. 8.14 is in fact the median of the values

$$x_1, \ldots, x_1, x_2, \ldots, x_2, \ldots, x_N, \ldots, x_N,$$

where each sample x_i is repeated γ_i times. The operation defined by Eq. 8.14 is exactly the **weighted median** of $x_1, \ldots, x_N$ with weights $\gamma_1, \ldots, \gamma_N$ and is denoted by

$$\mathrm{WM}\, [\gamma_1 \diamond x_1, \ldots, \gamma_N \diamond x_N]. \tag{8.15}$$

We saw that the median filter is the counterpart (in the median world) of the simple moving average. In the same way, the weighted median filter

$$y(n) = \mathrm{WM}\, [\gamma_{-k} \diamond x(n - k), \ldots, \gamma_k \diamond x(n + k)] \tag{8.16}$$

is the counterpart of the linear FIR filter

$$y(n) = \left(\frac{1}{\sum_{i=-k}^{k} \gamma_i} \right) \sum_{i=-k}^{k} \gamma_i x(n + i). \tag{8.17}$$

It has been shown [221] that weighted median filters can be succesfully used in many applications, especially in image processing. Compared to the standard median filter, where the only adjustable parameter is the window size, weighted median filters offer much more design freedom, as we can choose any positive weights. There are also effective methods to compute optimal weights adaptively. As we shall see later, even though there are an infinite number of weight combinations, there are only finitely many distinct weighted medians of a fixed window size.

There is a simple expression involving the defining positive Boolean function for the output distribution of a stack filter in the case of *i.i.d.* inputs [222]. Using this formula, certain statistical properties of stack and median-type filters can be reduced to the properties of Boolean functions. To derive these results in such a form that they also hold for weighted median filters over real-valued signals, we first define the continuous stack filter. Let $f(x_{-k}, \ldots, x_k)$ be a positive Boolean function (not identically =1) and $\mathbf{x} = \ldots, x(n-1), x(n), x(n+1), \ldots$ be a real-valued signal. Define a threshold function $T(x, \beta)$ by

$$T(x, \beta) = \begin{cases} 1 & \text{if } x \geq \beta; \\ 0 & \text{otherwise.} \end{cases} \tag{8.18}$$

Then the output of the continuous stack filter defined by f with input $\mathbf{x}$ at time instant n is

$$y(n) = \max \{\beta : f(T(x(n-k), \beta), \ldots, T(x(n+k), \beta)) = 1\}. \tag{8.19}$$

For example, if $f(x_{-1}, x_0, x_1) = x_{-1}x_0 + x_{-1}x_1 + x_0x_1$, we have just the three-point median for real-valued signals

We now investigate the random variable defined by the output of a continuous stack filter where the inputs are real-valued random variables. We shall consider the case when the input components (samples) of the input signal are identically distributed independent random variables. As we are only interested in the statistical properties of the output, we shall adopt the more standard indexing of the variables of the positive Boolean function. That is, we consider the positive Boolean function $f(\mathbf{x}) = f(x_1, \ldots, x_n)$. Let the inputs $X_1, \ldots, X_n$ be independent and identically distributed random variables with a common distribution function $\Phi(t)$ and define the random variable Y by

$$Y = \max \{\beta : f(T(X_1, \beta), \ldots, T(X_n, \beta)) = 1\}. \tag{8.20}$$

Now, the distribution function $\Psi(t)$ of Y can be written as

$$\Psi(t) = \sum_{i=0}^{n} A_i (1 - \Phi(t))^i \Phi(t)^{n-i}, \tag{8.21}$$

where the numbers A_i are defined by

$$A_i = |\{\mathbf{x} \in \{0, 1\}^n : f(\mathbf{x}) = 0, w_H(\mathbf{x}) = i\}|, \tag{8.22}$$

where $|S|$ means the cardinality of the set S and $w_H(\mathbf{x})$ denotes the number of 1's in $\mathbf{x}$, that is, its Hamming weight. To see this, note that

$$\Psi(t) = P\{Y \le t\}$$

and the input space can be divided into 2^n mutually exclusive events

$$(-\infty, t] \times (-\infty, t] \times \cdots \times (-\infty, t]$$

$$(t, \infty) \times (-\infty, t] \times \cdots \times (-\infty, t]$$

$$\vdots \tag{8.23}$$

$$(t, \infty) \times (t, \infty) \times \cdots \times (t, \infty).$$

A typical event having i terms of type (t, ∞) and $n - i$ terms of type $(-\infty, t]$ has probability $(1 - \Phi(t))^i \Phi(t)^{n-i}$. The event $\{Y \le t\}$ is the union of exactly those events in Eq. 8.23 whose terms of type (t, ∞) match with $1's$ in some $\mathbf{x} \in \{0, 1\}^n$ with $f(\mathbf{x}) = 0$. As the events in Eq. 8.23 are mutually exclusive, we can write the probability of the event $\{Y \le t\}$ as the sum in Eq. 8.21.

As an example, compute the output distribution function of the weighted median WM $[1 \diamond x(n-2), 2 \diamond x(n-1), 3 \diamond x(n), 2 \diamond x(n+1), 1 \diamond x(n+2)]$ when the input signal values are independent and identically distributed with a common distribution function $\Phi(t)$. Formula 8.21 shows that we need only compute the number of distinct Boolean vectors of each Hamming weight $i = 0, 1, 2, 3, 4, 5$ such that $f(\mathbf{x}) = 0$, where f is the linearly separable Boolean function

$$f(x_1, x_2, x_3, x_4, x_5) = \begin{cases} 1 & \text{if } x_1 + 2x_2 + 3x_3 + 2x_4 + x_5 > 5 \\ 0 & \text{otherwise.} \end{cases}$$

By examining the possibilities listing in Table 8.2 we see that the numbers A_i are $A_0 = 1$, $A_1 = 5$, $A_2 = 8$, $A_3 = 2$, $A_4 = A_5 = 0$. Thus,

$$\Psi(t) = \Phi(t)^5 + 5(1 - \Phi(t))\Phi(t)^4 + 8(1 - \Phi(t))^2\Phi(t)^3 + 2(1 - \Phi(t))^3\Phi(t)^2$$

$$= 2\Phi(t)^2 + 2\Phi(t)^3 - 5\Phi(t)^4 + 2\Phi(t)^5.$$

x					w	f(x)	x					w	f(x)
0	0	0	0	0	0	0	1	0	0	0	0	1	0
0	0	0	0	1	1	0	1	0	0	0	1	2	0
0	0	0	1	0	1	0	1	0	0	1	0	2	0
0	0	0	1	1	2	0	1	0	0	1	1	3	0
0	0	1	0	0	1	0	1	0	1	0	0	2	0
0	0	1	0	1	2	0	1	0	1	0	1	3	1
0	0	1	1	0	2	1	1	0	1	1	0	3	1
0	0	1	1	1	3	1	1	0	1	1	1	4	1
0	1	0	0	0	1	0	1	1	0	0	0	2	0
0	1	0	0	1	2	0	1	1	0	0	1	3	0
0	1	0	1	0	2	0	1	1	0	1	0	3	1
0	1	0	1	1	3	1	1	1	0	1	1	4	1
0	1	1	0	0	2	1	1	1	1	0	0	3	1
0	1	1	0	1	3	1	1	1	1	0	1	4	1
0	1	1	1	0	3	1	1	1	1	1	0	4	1
0	1	1	1	1	4	1	1	1	1	1	1	5	1

Table 8.2. The elements of $\{0,1\}^5$.

In the actual computation of the numbers A_i for a weighted median it is not necessary to list the vectors $\mathbf{x} \in \{0,1\}^n$ for which $f(\mathbf{x}) = 0$. The computation can be done extremely fast using generating functions [5]. Consider the weighted median WM $[w_1 \diamond x_1, \ldots, w_n \diamond x_n]$, where $\sum_{i=0}^n w_i = 2T + 1$. We form the polynomial

$$P(\xi, \eta) = \prod_{i=1}^n (1 + \xi \eta^{w_i}),$$

(8.24)

which can be written in the form

$$
\begin{aligned}
P(\xi, \eta) &= \sum_{\mathbf{x} \in \{0,1\}^n} \prod_{i=1}^n (\xi \eta^{w_i})^{x_i} \\
&= \sum_{\mathbf{x} \in \{0,1\}^n} \xi^{\sum x_i} \eta^{\sum w_i x_i}.
\end{aligned}
$$

(8.25)

From Eq. 8.25 we see that $P(\xi, \eta)$ has one term for each $\mathbf{x} \in \{0,1\}^n$, the exponent of ξ is the Hamming weight, and the exponent of η is $\sum w_i x_i$. Deleting all terms whose power of η is at least $T + 1$ we obtain exactly the vectors $\mathbf{x}$ for which WM $[w_1 \diamond x_1, \ldots, w_n \diamond x_n] = 0$. Thus, the procedure for computing A_i can be written as follows:

1. Form $P(\xi, \eta) = \prod_{i=1}^{n}(1 + \xi\eta^{w_i})$.

2. Expand $P(\xi, \eta) = \sum \xi^{\sum x_i} \eta^{\sum w_i x_i}$.

3. Collect the powers of η: $P(\xi, \eta) = \sum_{k=0}^{2T+1} S_k(\xi)\eta^k$.

4. Truncate with respect to η at T: $Q(\xi, \eta) = \sum_{k=0}^{T} S_k(\xi)\eta^k$.

5. Now, $R(\xi) = Q(\xi, 1) = \sum_{i=0}^{n} A_i\xi^i$.

Notice that once we have the generating function

$$R(\xi) = \sum_{i=0}^{n} A_i\xi^i, \tag{8.26}$$

by Eq. 8.21, the output distribution $\Psi(t)$ can be written simply as

$$\Psi(t) = \Phi(t)^n R\left(\frac{1 - \Phi(t)}{\Phi(t)}\right). \tag{8.27}$$

To illustrate the method, we compute the output distribution of WM $[1 \diamond x_1, 2 \diamond x_2, 3 \diamond x_3, 2 \diamond x_4, 1 \diamond x_5]$ using generating functions. Now,

$$\begin{aligned}
P(\xi, \eta) &= (1 + \xi\eta)^2(1 + \xi\eta^2)^2(1 + \xi\eta^3) \\
&= 1 + 2\xi\eta + (2\xi + \xi^2)\eta^2 + (\xi + 4\xi^2)\eta^3 \\
&\quad + (3\xi^2 + 2\xi^3)\eta^4 + (2\xi^2 + 3\xi^3)\eta^5 + (4\xi^3 + \xi^4)\eta^6 \\
&\quad + (\xi^3 + 2\xi^4)\eta^7 + 2\xi^4\eta^8 + \xi^5\eta^9,
\end{aligned}$$

and

$$Q(\xi, \eta) = 1 + 2\xi\eta + (2\xi + \xi^2)\eta^2 + (\xi + 4\xi^2)\eta^3 + (3\xi^2 + 2\xi^3)\eta^4.$$

Thus, the generating function for A_i is

$$R(\xi) = 1 + 5\xi + 8\xi^2 + 2\xi^3$$

giving

$$\Psi(t) = \Phi(t)^n R\left(\frac{1 - \Phi(t)}{\Phi(t)}\right)$$

$$= 2\Phi(t)^2 + 2\Phi(t)^3 - 5\Phi(t)^4 + 2\Phi(t)^5.$$

In Fig. 8.4 the noise attenuation capability of this filter is illustrated by plotting the input and output densities in the case of Cauchy distributed input.

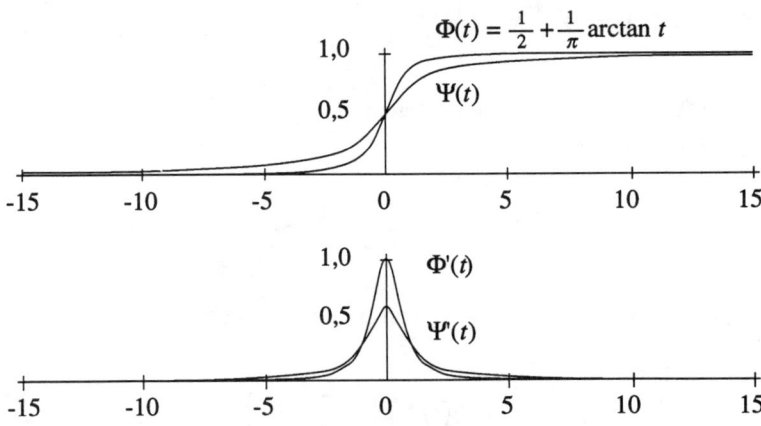

Fig. 8.4. The noise attenuation of the weighted median filter $y(n) = \text{WM}[x(n - 2), 2 \diamond x(n-1), 3 \diamond x(n), 2 \diamond x(n+1), x(n+2)]$ **for Cauchy distributed input.** Φ **is the input distribution and** Ψ **is the output distribution,** Φ' **and** Ψ' **are the corresponding densities.**

Chapter 9

Optimization of Weighted Median Filters

There has been much work on how to find the best stack or weighted-order statistic filter for a particular application. In the structural approach one defines beforehand what shapes or signal forms must be eliminated and what must be retained and tries to build a filter satisfying these criteria. This may be satisfactory for an isolated case but, in general, produces unpredictable algorithms. A more sound approach is to use estimation theory in the sense that the goal is to find the filter that best estimates the desired signal given a corrupted version of the signal as the input. A natural measure of the value of the filter is the mean absolute error (MAE) between the actual output of the filter and the desired output. If the signal and the noise are jointly stationary, using linear programming an optimal stack filter can be found. However, there are severe limitations: the window size must be small because the size of the linear programming problem is exponential with respect to the window size, and, real images very seldom fulfill the requirement of joint stationarity. Moreover, it is necessary to compute or estimate the coefficients of the cost function to be minimized. The problem of modeling can be partly avoided by using adaptive methods [107,221]. However, the large number of "free" parameters in stack filters of reasonable window size makes convergence very slow. A practical solution is to limit the number of parameters by restricting the search into a subclass of stack filters. If the subclass is well chosen, the resulting suboptimal filter is satisfactory. In Ref. [220] an adaptive method for optimal weighted order statistic filtering in image processing was developed. In the next chapter we consider optimization of binary nonlinear filters.

In the following we briefly describe a new approach to designing optimal weighted median filters. As noted above, full optimization of a stack filter under arbitrary conditions is, computationally, next to impossible due to the large number of parameters. We consider the following restricted problem, which is general enough to give meaningful filters but computationally manageable. The idea is to find the weighted median filter that, first, leaves a set of signals unchanged and, on the other hand, attenuates white noise as effectively as possible.

Let $f(\mathbf{x})$ be a self dual positive Boolean function; i.e., $f(\mathbf{x}) = \overline{f(\overline{\mathbf{x}})}$ for all $\mathbf{x} \in \{0,1\}^n$.

Denote

$$M_i = |\{\mathbf{x} \in \{0,1\}^n \mid f(\mathbf{x}) = 1, \ w(\mathbf{x}) = i\}|, \qquad (9.1)$$

where $M_i = \binom{n}{i} - A_i$, $i = 0, \ldots, n$, and A_i, $w(\mathbf{x})$ are as in Eq. 8.22. The following theorem [222] is the basis for the method.

Theorem 9.1. *Let the inputs of a self dual stack filter with window size $N = 2k + 1$ be independent and identically distributed with a common distribution function $\Phi(t)$. The output distribution of the stack filter has distribution function*

$$\Psi(t) = \Psi_m(t) + \sum_{i=1}^{k} M_i \left(\Phi(t)^i (1 - \Phi(t))^{N-i} - \Phi(t)^{N-i} (1 - \Phi(t))^i \right), \qquad (9.2)$$

where $\Psi_m(t)$ is the distribution function of the median.

From Eq. 9.2 we see that, as estimators of the location parameter of symmetric distributions, self dual stack filters are unbiased. Thus, if m_x denotes the expectation of the inputs and $\Phi(t)$ is symmetric with respect to m_x, that is,

$$\Phi(m_x - t) + \Phi(m_x + t) = 1 \qquad (9.3)$$

then for the expectation $E[y]$ of the output of a self dual stack filter we have

$$E[y] = m_x.$$

A more important consequence of Eq. 9.2 is that we can, for symmetric distributions, express an output moment of a self dual stack filter as a sum of the corresponding moment for median filter and a quantity depending only on Φ and the coefficients M_i [220].

Theorem 9.2. *Consider a self dual stack filter having window size $N = 2k + 1$ and let the input be i.i.d. random variables have a common distribution function Φ, symmetric with respect to the origin. Then the rth order output moment of the stack filter is*

$$\mu_r = \mu_r^{(m)} + \sum_{i=1}^{k} M_i L_i(N, \Phi, r), \qquad (9.4)$$

where $\mu_r^{(m)}$ is the rth order central moment of the median filter of size N and

$$L_i(N,\Phi,r) = \int\limits_{-\infty}^{\infty} |t|^r \left[\frac{d}{dt}\left(\Phi(t)^i(1-\Phi(t))^{N-i} - \Phi(t)^{N-i}(1-\Phi(t))^i \right) \right] dt. \quad (9.5)$$

As one can show that

$$L_i(N,\Phi,r) \geq 0, \quad (9.6)$$

we can see that, of all self dual stack filters of fixed size, the median filter has the best noise attenuation [220]. In the following we use Eq. 9.4 to find among the weighted median filters satisfying certain structural constraints the one that will minimize the output variance.

Consider a weighted median filter with window size N, positive integer weights $w_1,\ldots,w_N$, and suppose that $\sum_{i=1}^N w_i = 2T+1$. Let $f(x_1,\ldots,x_N)$ be the self dual positive Boolean function defined by

$$f(x_1,\ldots,x_N) = 1 \quad \Longleftrightarrow \quad \sum_{i=1}^N w_i x_i \geq T, \quad (9.7)$$

where on the right-hand side of Eq. 9.7 the binary variables x_i are interpreted as real 0 and 1. We want to have a stack filter that will retain and eliminate prescribed image details, which means that some a priori set of signals must have output 0 and some other a priori set of signals must have output 1. This leads to a set of equations that f must satisfy:

$$\begin{cases} f\left(\mathbf{x}^{(1)}\right) = 0 \\ \quad\vdots \\ f\left(\mathbf{x}^{(l)}\right) = 0 \\ f\left(\mathbf{x}^{(l+1)}\right) = 1 \\ \quad\vdots \\ f\left(\mathbf{x}^{(h)}\right) = 1, \end{cases} \quad (9.8)$$

where $\mathbf{x}^{(i)} \in \{0,1\}^N$, $i = 1,\ldots,h$. Equations 9.8 give rise to the following inequalities for the weights,

$$\begin{cases} x_1^{(1)}w_1 + \ldots + x_N^{(1)}w_N \leq T \\ \quad\vdots \\ x_1^{(l)}w_1 + \ldots + x_N^{(l)}w_N \leq T \\ x_1^{(l+1)}w_1 + \ldots + x_N^{(l+1)}w_N \geq T+1 \\ \quad\vdots \\ x_1^{(h)}w_1 + \ldots + x_N^{(h)}w_N \geq T+1 \\ w_1 + \ldots + w_N = 2T+1. \end{cases} \tag{9.9}$$

The optimization problem can now be formulated as follows:

(i) Compute the coefficients $L_i(N, \Phi, 2)$, $i = 1, \ldots, k - \frac{N-1}{2}$.

(ii) Minimize $\sum_{i=1}^{k} M_i L_i$ under Eq. 9.9.

Notice that because the coefficients M_i depend on w_i in a nonlinear manner, this is a nonlinear optimization problem.

Let us apply the above method to find the optimal centrally symmetric weighted median filter that can preserve pulses of length 2. Assume that the weight vector is given by

$$\mathbf{w} = (w_3, w_2, w_1, w_0, w_1, w_2, w_3). \tag{9.10}$$

The requirement that length two pulses are preserved leads to

$$2(w_0 + w_1) > w_0 + 2w_1 + 2w_2 + 2w_3 \tag{9.11}$$

or

$$w_0 > 2(w_2 + w_3). \tag{9.12}$$

With lengthy but straightforward computations one can show that $\mathbf{w} = (1,1,3,5,3,1,1)$ yields coefficients M_1^*, M_2^*, M_3^* such that $M_1 \geq M_1^*, M_2 \geq M_2^*, M_3 \geq M_3^*$ for any other choice of $\mathbf{w}$. This means by Eqs. 9.4 and 9.6 that $\mathbf{w} = (1,1,3,5,3,1,1)$ minimizes output noise for any symmetric noise distribution. Figure 9.1 illustrates the behavior of optimal weighted median filters designed using this method.

Fig. 9.1. This figure illustrates the behavior of weighted median filters that have been optimized under structural constraints. (a) Original image.

**Fig. 9.1. (b) Original plus impulsive noise; the probability for white pixels is 6%
and the probability for black pixels is 6%.**

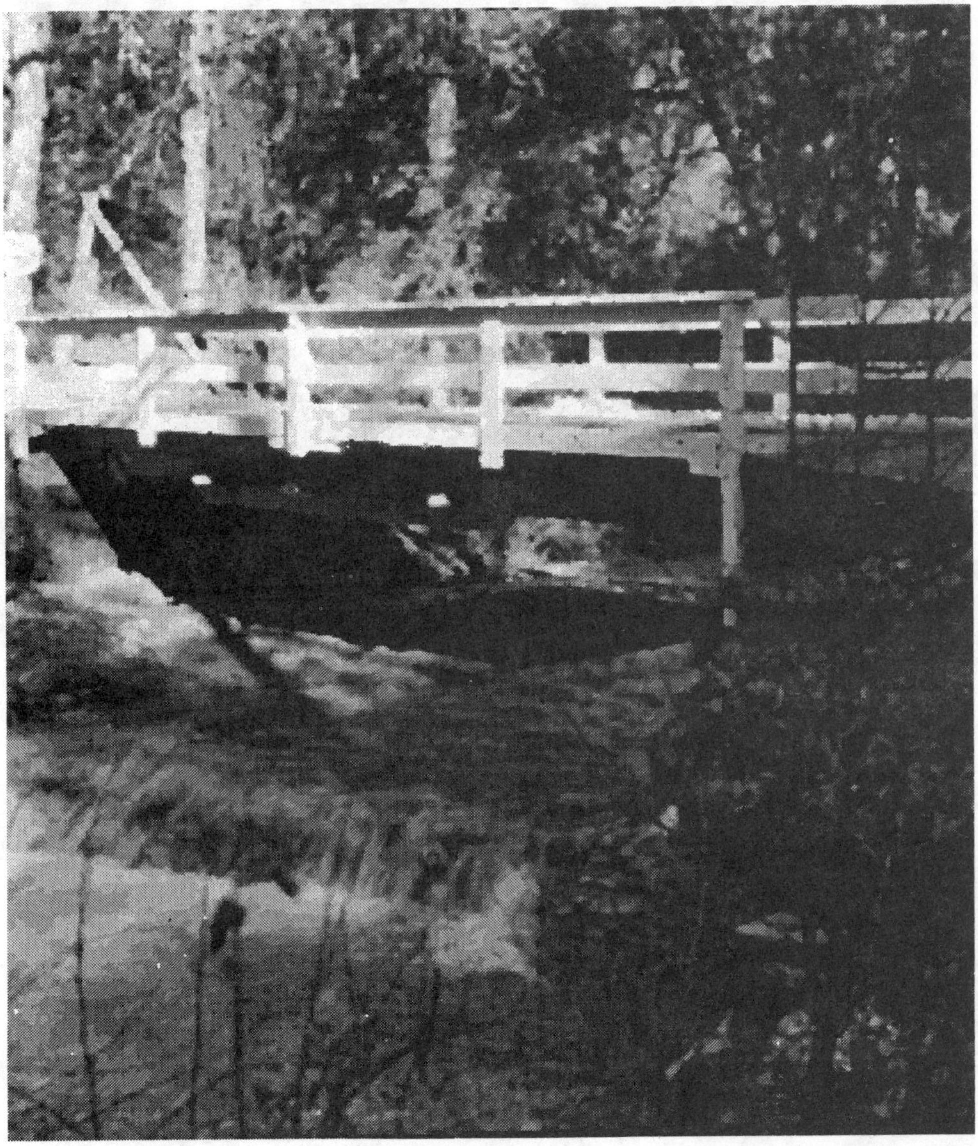

Fig. 9.1. (c) Noisy image filtered with the standard 3×3 median, mask

1	1	1
1	1	1
1	1	1

.

Fig. 9.1. (d) Noisy image filtered with the optimal corner-preserving weighted median filter, mask
$$\begin{array}{|c|c|c|} \hline 2 & 3 & 2 \\ \hline 3 & 5 & 3 \\ \hline 2 & 3 & 2 \\ \hline \end{array}$$
.

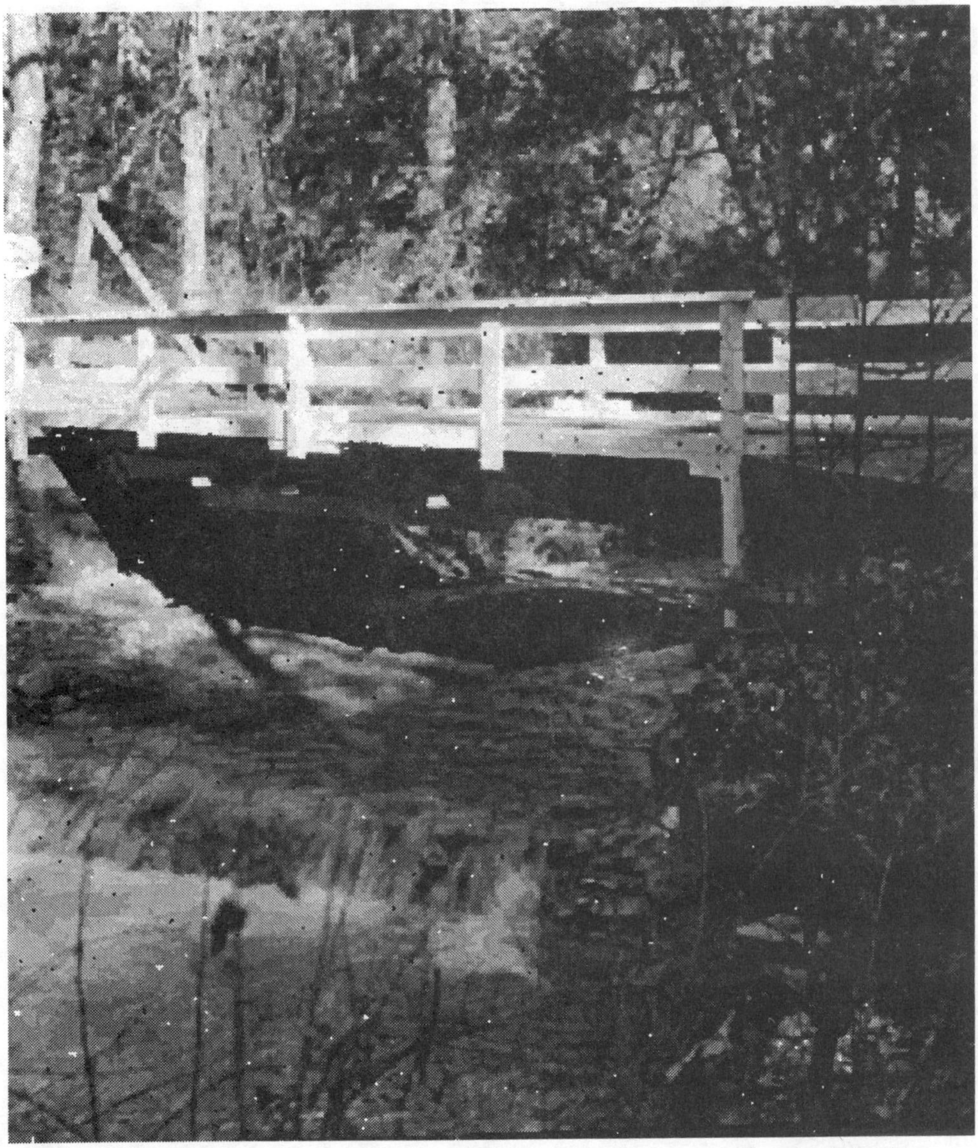

Fig. 9.1. (e) Noisy image filtered with the optimal weighted median that preserves horizontal and vertical lines, mask $\begin{array}{|c|c|c|}\hline 1 & 3 & 1 \\ \hline 3 & 5 & 3 \\ \hline 1 & 3 & 1 \\ \hline \end{array}$ **.**

Fig. 9.1. (f) Noisy image filtered with the optimal weighted median that preserves horizontal and vertical lines and also corners, mask

$$\begin{bmatrix} 1 & 2 & 1 \\ 2 & 5 & 2 \\ 1 & 2 & 1 \end{bmatrix}$$

.

Fig. 9.1. (g) Noisy image filtered with the optimal weighted median filter that preserves horizontal, vertical, and diagonal lines, mask

1	1	1
1	5	1
1	1	1

.

Chapter 10

Optimal Binary Filters

From the perspective of statistical estimation, the filtering problem involves operating on a collection of random observations in order to estimate some unobserved value. In image processing, an image is observed and we would like to operate on it so that the resulting filtered image is a good estimate of the desired image. In the language of restoration, we observe a noisy image and filter it to produce a restored image close to the ideal image. Implicit in the formulation of the problem is the existence of a goodness criterion by which to measure the closeness of the filtered and ideal images. In defining a goodness criterion we take into account that the images are modeled as random processes and we recognize that restoration is relative to the stated goodness criterion. An optimal filter is one that, among some class of filters, achieves the least error as measured by the goodness criterion. In the present chapter we consider the design of optimal translation-invariant binary filters, both increasing and nonincreasing. Given the representation of these types of filters in terms of erosions and hit-or-miss transforms, it is natural to expect that optimal filter design will be in the context of mathematical morphology.

10.1. Optimal Mean-Absolute-Error Filtering

We begin with a brief review of optimal finite-observation filters in the framework of the mean-absolute-error criterion. Given n observation random variables $X_1, X_2, \ldots, X_n$, which we interpret as a random vector $\mathbf{X} = (X_1, X_2, \ldots, X_n)$, and the random variable Y to be estimated, the optimal mean-absolute-error (MAE) estimator is the function $g(\mathbf{X}) = g(X_1, X_2, \ldots, X_n)$ that mimimizes the expected value (mean-absolute error)

$$\text{MAE}\,\langle g \rangle = E[\,|Y - g(X_1, X_2, \ldots, X_n)|\,]. \tag{10.1}$$

It is well known that in the binary setting, where mean-absolute error and mean-square error are identical, the optimal MAE estimator is given by the conditional expectation,

$$E[\,Y \mid X_1, X_2, \ldots, X_n\,] = P(Y = 1 \mid X_1, X_2, \ldots, X_n). \tag{10.2}$$

Since we are in a binary setting, the conditional expectation involves rounding. Thus, Eq. 10.2 is interpreted to mean

$$E[Y \mid \mathbf{X}] = \begin{cases} 1, & \text{if } P(Y = 1 \mid \mathbf{X}) > 0.5 \\ 0, & \text{if } P(Y = 1 \mid \mathbf{X}) \le 0.5. \end{cases} \tag{10.3}$$

Rather than find the best estimator, it is common to look for the best estimator among a collection of estimators, thereby restricting the nature of the estimation rule g. Historically, for gray-scale filtering, the most commonly employed filter has been the optimal linear mean-square-error estimator, for which g is restricted to linear combinations of the observation variables. Here we study optimal MAE digital binary filtering within the context of general MAE optimal estimation. Since linearity is not relevant to the binary setting, the optimal filters will, ipso facto, be nonlinear. Key to the study is the placement of morphological operators into the standard statistical estimation framework.

We will be considering the filtering of random binary images (sets). There is an observation set process $\mathbf{S}$ and an ideal set process $\mathbf{S}_0$. Any particular observed image S is a realization of the process $\mathbf{S}$ and any particular ideal image S_0 is a realization of $\mathbf{S}_0$. In practice we observe a single realization of $\mathbf{S}$ and predict the most likely corresponding realization of $\mathbf{S}_0$.

10.2. Optimal Erosions

We first consider estimation by a single erosion. The problem is posed as follows: Given that we observe a random image (set) $\mathbf{S}$ in some window W_x at pixel x, how do we pick the structuring element A so that $A + x$ is a subset of W_x and the random $\{0, 1\}$-valued estimator defined by erosion is an optimal estimator of the random variable Y? Referring to Fig. 10.1, we can visualize the problem as one of restoration. The observed image $\mathbf{S}$ is a degraded version of the ideal image $\mathbf{S}_0$ (both random), we observe the values $X_1, X_2, \ldots, X_n$ of $\mathbf{S}$ in the neighborhood W_x of the pixel x, Y is the ideal value at the pixel x, and we wish to choose the structuring element A so that erosion of $\mathbf{S}$ by A provides optimal estimation (prediction) of Y. Parts (a) and (b) of the figure show realizations of the process $\mathbf{S}$ that, respectively, contain and do not contain $A + x$. Thus, for Fig. 10.1, the prediction of the corresponding random variable Y would be 1 for the realization of part (a) and 0 for the realization of part (b).

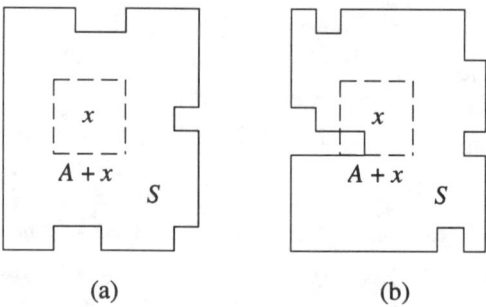

(a)   (b)

Fig. 10.1. Erosion estimation: (a) realization containing structuring element; (b) realization not containing structuring element.

To intepret the preceding geometric considerations in a manner suitable for statistical analysis, we consider the random binary values of the random set lying in the window to compose a random vector $\mathbf{X} = (X_1, X_2, \ldots, X_n)$. Each realization of $\mathbf{X}$ is a deterministic n-vector $\mathbf{x} = (x_1, x_2, \ldots, x_n)$, where $x_k = 1$ if the kth pixel in the window lies in the corresponding realization of the random set and $x_k = 0$ if it does not. Using the $0 - 1$ image representation, a structuring element A lying within an n-point window is viewed as an n-vector $A = (a_1, a_2, \ldots, a_n)$ of zeros and ones, and each realization $\mathbf{x}$ of the windowed image process provides a binary value

$$\mathbf{x} \ominus A = (x_1, x_2, \ldots, x_n) \ominus (a_1, a_2, \ldots, a_n)$$
$$= \min\{x_k : a_k = 1\}. \tag{10.4}$$

The salient point is that now, for the binary random vector $\mathbf{X}$ and the fixed structuring element A,

$$\mathbf{X} \ominus A = \min\{X_k : a_k = 1\} \tag{10.5}$$

is a random variable and as such can be treated as an estimator of another random variable Y. In terms of the fundamental MAE expression of Eq. 10.1, the estimation rule g takes the form $g(\mathbf{X}) = \mathbf{X} \ominus A$. The optimal erosion filter is the one defined by the structuring element A that minimizes

$$\begin{aligned} \text{MAE}\,\langle A \rangle &= E[|Y - \mathbf{X} \ominus A|] \\ &= P(|Y - \mathbf{X} \ominus A| = 1) \\ &= P(Y \neq \mathbf{X} \ominus A) \\ &= \sum \{f(x_1 x_2, \ldots, x_n, y) : y \neq \mathbf{x} \ominus A\}, \end{aligned} \tag{10.6}$$

where

$$f(x_1, x_2, \ldots, x_n, y) = P(X_1 = x_1, X_2 = x_2, \ldots, X_n = x_n, Y = y) \qquad (10.7)$$

is the joint probability density for the random $(n + 1)$-vector $(X_1, X_2, \ldots, X_n, Y)$. A key point is that the error is found by summing up the probabilities (over all realizations) that erosion by A gives the incorrect value for Y.

It should be recognized that the entire foregoing development considered estimation at a single pixel. A natural question arises: Does optimization yield a different structuring element at each pixel (which would mean that we do not have an erosion in the ordinary sense)? No – so long as we assume the joint process is stationary. This is the kind of assumption we make in practice and it is analogous to the assumption of wide-sense stationarity made in the application of spatially invariant linear filters. For many binary image processes, especially text, the assumption leads to good results.

In practice, MAE $\langle A \rangle$ is estimated from realizations of both the ideal and observation processes. In the case of restoration, we take an ideal realization S_0 and degrade it (by whatever noise process is under study) to produce a corresponding realization S of the noisy process. We then erode S by A and compare $S \ominus A$ to S_0 pixelwise. An estimate of MAE $\langle A \rangle$ is obtained by dividing the numer of pixels at which $S \ominus A$ and S_0 disagree by the total number of pixels considered. Such estimation from realizations provides, in the usual statistical sense, an estimate of MAE $\langle A \rangle$ whose precision increases with the number of pixels observed.

10.3. Optimal Increasing Filters via the Matheron Representation

It is unlikely that a single erosion, even if optimal, will provide good filtering, and therefore extension to more general filters is necessary. For design of increasing filters, the Matheron representation is key [33,34]. To adapt the representation to optimization, we recognize a limitation to n-point erosions, since the statistical analysis is based on n observations. For a single erosion, the optimal structuring element is limited to being a subset of the n points at which the observations are made. Using the Matheron representation as our guide and noting that maximum plays the role of union when sets are given

$0-1$ representation, we define an n-**observation morphological filter** to be a functional of the form

$$\Psi(\mathbf{x}) = \max_i \{\mathbf{x} \ominus B_i\}$$

$$= \max_i \{\min_j \{x_j : b_{ij} = 1\}\}, \tag{10.8}$$

where $\mathbf{x} = (x_1, x_2, \ldots, x_n)$ and $B_i = (b_{i1}, b_{i2}, \ldots, b_{in})$ are deterministic binary n-vectors. The collection $\{B_i\}$ is called the **kernel** of Ψ and, if the collection $\{B_i\}$ is minimal, then it is called the **basis** for Ψ and denoted by Bas $[\Psi]$.

Extension of the optimal filter solution to general increasing binary filters involves minimizing the mean-absolute error

$$\text{MAE} \langle \Psi \rangle = E[|Y - \Psi(X_1, X_2, \ldots, X_n)|] \tag{10.9}$$

over all possible n-observation increasing filters Ψ. Since Ψ is fully determined by its basis, finding the optimal n-observation filter reduces to selecting the subset of the 2^n binary structuring elements (including the null one) that yields minimum MAE $\langle \Psi \rangle$. Owing to basis minimality, many of these subsets can be eliminated from consideration.

This elimination can be described by means of a graph, which we call the **basis graph**. Write each structuring element as a string of ones and zeros and use these strings to construct a graph with n horizontal levels. Level 0 (the top level) consists of the single string of all ones, level 1 consists of the $n = C(n, 1)$ strings with $n-1$ ones and a single zero, level 2 consists of the $C(n, 2)$ strings with $n-2$ ones and two zeros,..., level $n-1$ consists of the $n = C(n, n-1)$ strings with a single one and $n-1$ zeros, and level n consists of the null element. Arrows in the basis graph are defined by basis minimality. If structuring elements $A = (a_1, a_2, \ldots, a_n)$ and $B = (b_1, b_2, \ldots, b_n)$ are such that $b_k = 1$ implies $a_k = 1$, then, not only does A lie on a higher tier of the basis graph than does B, but if B lies in Bas $[\Psi]$, then A cannot, since $B \leq A$. Thus, the number of possible bases, and therefore the number of possible optimal filters, is reduced. The basis graph is completed by letting an arrow point from vertex B to vertex A if $B \leq A$. A collection of nodes is a basis if there does not exist a path from any node in the collection to any other node in the collection. The basis graph for $n = 4$ observations is shown in Fig. 10.2.

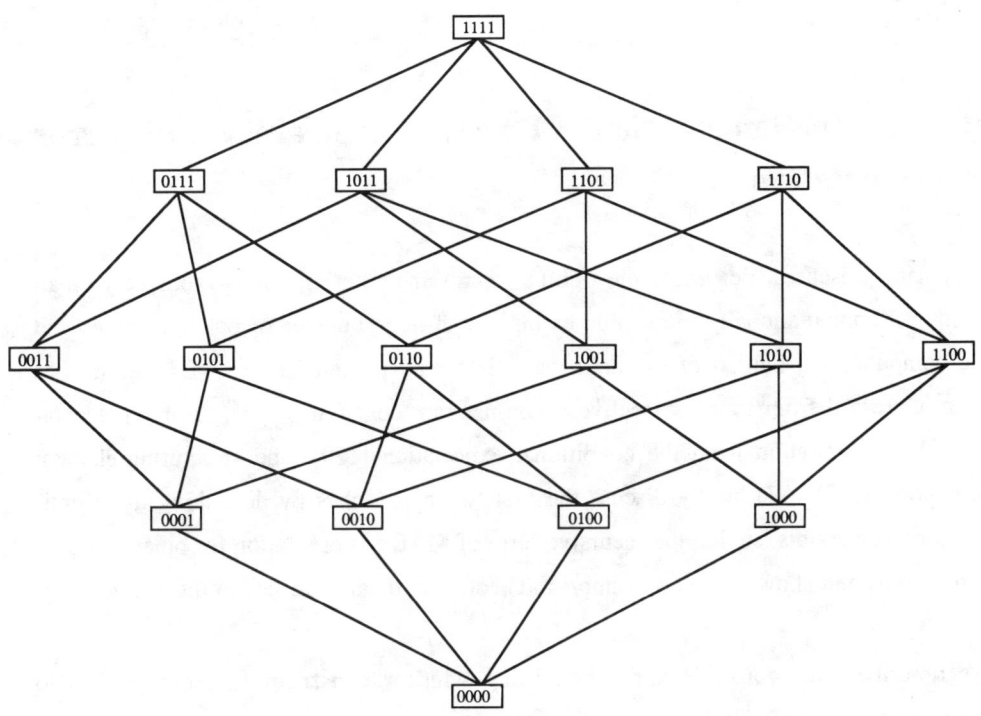

Fig. 10.2. Basis graph for four observations.

The optimal increasing MAE filter is found by searching the basis graph for all possible bases, for each basis constructing the corresponding filter via the Matheron representation, and selecting Bas $[\Psi]$ to minimize MAE $\langle \Psi \rangle$.

To express MAE $\langle \Psi \rangle$, suppose Ψ has a basis consisting of m structuring elements. Then Ψ is given by the maximum

$$\Psi(\mathbf{X}) = [\mathbf{X} \ominus B_1] \vee [\mathbf{X} \ominus B_2] \vee \ldots \vee [\mathbf{X} \ominus B_m] \tag{10.10}$$

and

$$\text{MAE} \langle \Psi \rangle = \sum \{ f(x_1, x_2, \ldots, x_n, y) : y \neq \max \{ \mathbf{x} \ominus B_i \} \}. \tag{10.11}$$

If $\mathbf{B} = \{ B_1, B_2, \ldots, B_m \}$ is the basis for Ψ, we often write MAE $\langle \mathbf{B} \rangle$ in lieu of MAE $\langle \Psi \rangle$.

10.4.   Implementation: Design Constraints and Error Representation

As so far presented, design of the optimal increasing filter is, for any but very small windows, computationally intractable owing to the large number of potential bases and the computational burden of estimating MAEs for all potential bases. Tractability has been addressed from four perspectives: optimization constraint [111], error representation [110], derivation from the conditional expectation [129], and structuring-element adaptation [152]. We briefly discuss the first two approaches by describing three optimization constraints, stating the recursive form of MAE representation for binary erosion expansions, and showing some examples. Theoretical details are left to the literature.

We describe three constraints, the first being **window constraint**. If our goal is to optimize over a 5×5 window, we might instead optimize over a smaller subwindow. For instance, the window of Fig. 10.3 has proven useful for character restoration. Whereas the full 5×5 window contains 25 pixels and therefore has 2^{25} subsets, the subwindow contains 17 pixels and therefore has only 2^{17} subsets. Geometrically, the subwindow disallows certain nonsymmetric structuring elements. From a strictly logical perspective, window constraint need not be considered as a constraint because it simply results in optimization over a smaller window; however, from the perspective that we would like to optimize over the full 5×5 window, it does represent a constraint on optimality.

Fig. 10.3. Constrained window.

A second constraint commonly employed is **basis-size constraint**. Rather than necessarily finding the optimal basis, we find the optimal basis containing some number m of structuring elements. The number of structuring elements employed might be determined prior to the design search or it might be determined dynamically by plotting the MAE resulting from the optimal basis against the number of structuring elements in the basis. When the decrease in MAE for increasing the number of structuring elements becomes less significant (the MAE versus m curve flattens), design is terminated and the currently found basis is employed in a suboptimal fashion. Practice has shown this approach to be quite fruitful (see [44] for a theoretical characterization). A point to be kept in mind, and one that adversely impacts our ability to achieve fast design, is that the optimal m-element basis need not be a subcollection of the optimal $(m + 1)$-element basis; that is, there is no decreasing monotonicity with increasing basis size.

Last, but not least, we discuss **library constraint**. Rather than attempt to select the optimal basis from among all possible bases, library constraint imposes suboptimality by only selecting bases from some library of structuring elements. Libraries have been formed in a number of manners, including statistically [111]; however, here we discuss only **expert libraries**. These are formed from structuring elements known to be useful in filtering images from some observeration-image model to obtain images in some ideal-image model. For instance, since medians are useful for restoring certain types of images degraded by salt-and-pepper noise, structuring elements contained in median bases might be included in the library. If degradation has been caused by salt noise alone, then, since closings tend to correct salt-noise degradation, closing-basis structuring elements might be included in the library (see Fig. 10.4, where such structuring elements are called **hole fillers**). Certain types of images degraded by union noise tend to be restored by openings, so therefore opening-basis structuring elements might be included in the library. Generally, a library consists of various types of structuring elements and it is from among these that an optimal (suboptimal) filter is formed.

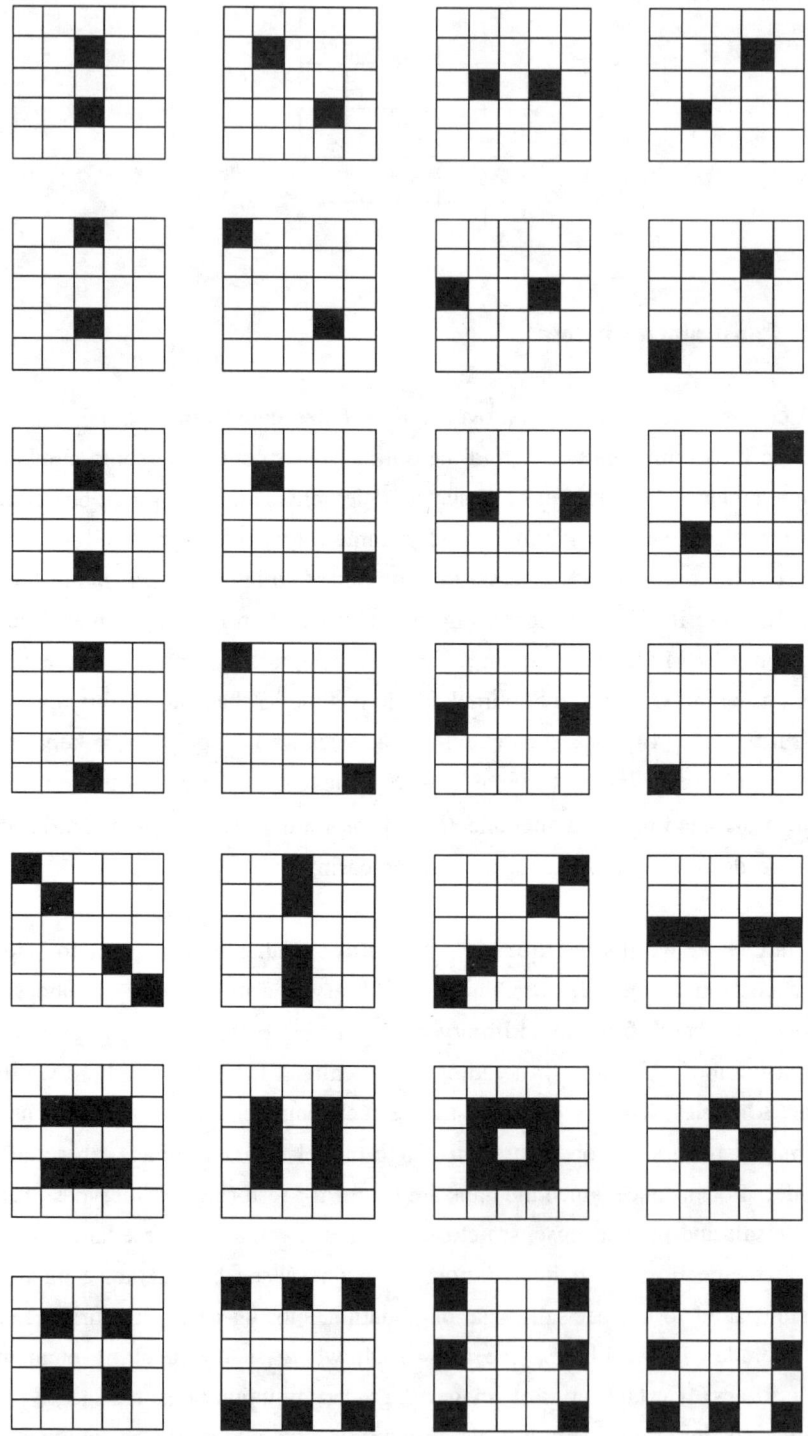

Fig. 10.4. Examples of hole fillers.

As described by Eq. 10.11, MAE associated with a basis **B** composed of m elements is found via realizations. Thus, finding minimal MAE appears to require realization error analysis for all bases of all sizes in question. Fortunately, such is not the case; in fact, a fundamental theorem regarding MAE for increasing filters represented via erosions shows that MAE for any basis can be expressed in terms of single-erosion MAEs, thereby relieving us of the need to perform error estimation for multiple-erosion filters. Moreover, the error representation possesses a recursive formulation that gives the error for an m-erosion filter in terms of the errors of two $(m-1)$-erosion filters and one single-erosion filter. Hence, errors for filters possessing increasing numbers of erosions can be found recursively from previously calculated errors. We state only the recursive form of the theorem [110]: An m-erosion binary increasing filter Ψ_m possessing basis Bas $[\Psi_m] = \{B_1, B_2, \ldots, B_m\}$ possesses MAE given by

$$\text{MAE}\langle\Psi_m\rangle = \text{MAE}\langle\Psi_{m-1}\rangle + \text{MAE}\langle B_m\rangle - \text{MAE}\langle\Phi_{m-1}\rangle, \tag{10.12}$$

where

$$\text{Bas}\langle\Psi_{m-1}\rangle = \{B_1, B_2, \ldots, B_{m-1}\} \tag{10.13}$$

$$\text{Bas}\langle\Phi_{m-1}\rangle = \{B_1 \cup B_m, B_2 \cup B_m, \ldots, B_{m-1} \cup B_m\}. \tag{10.14}$$

A filter derived via the foregoing optimization procedure has already been given in Chapter 4, the basis of the filter being provided in Fig. 4.5 and the original, noisy, and restored text images being shown in Figs. 4.2, 4.3, and 4.4, respectively. For a second, more detailed, example, consider the original text image of Fig. 10.5 and the noisy image of Fig. 10.6 that has suffered from both thinning and random salt noise. A library involving the constrained window of Fig. 10.3 has been derived from the image-noise model to find the optimal 8-erosion filter. Figures 10.7, 10.8, and 10.9 show the curve of MAE versus basis size for the optimal m-element filter, the optimal 8-erosion basis found from realizations, and the restored image obtained using this basis. In this case each structuring element consists of a single pixel, so that each yields an erosion that is just a simple translation. Thus, the union-of-erosion filter from the library is just dilation by the 3×3 origin-centered square absent the origin.

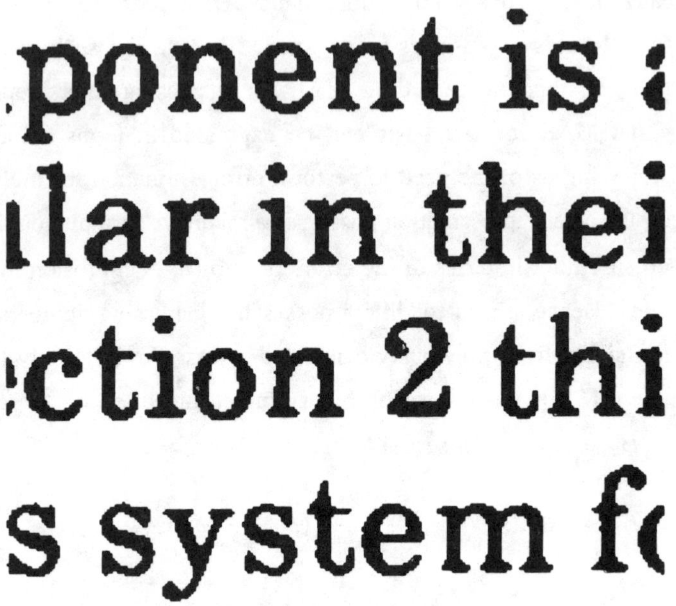

ponent is ﬂ

lar in thei

ction 2 thi

s system f(

Fig. 10.5. Original text image.

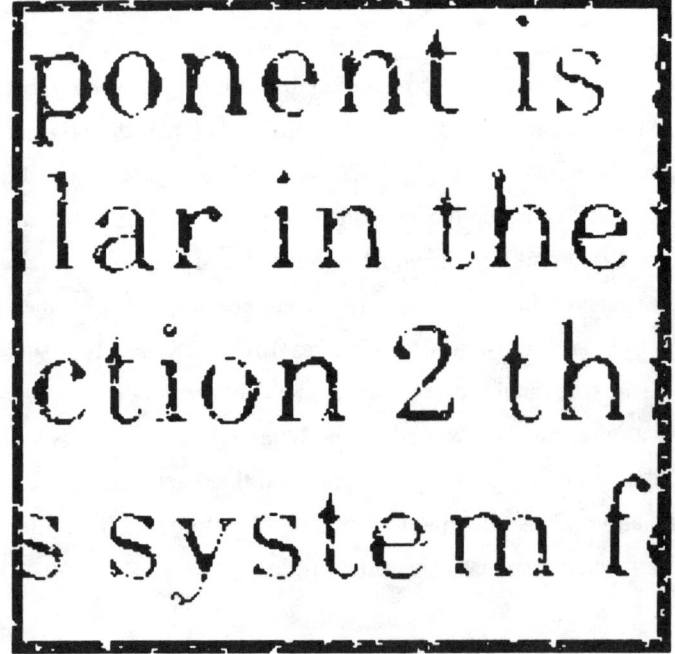

Fig. 10.6. Text image degraded by thinning and salt noise.

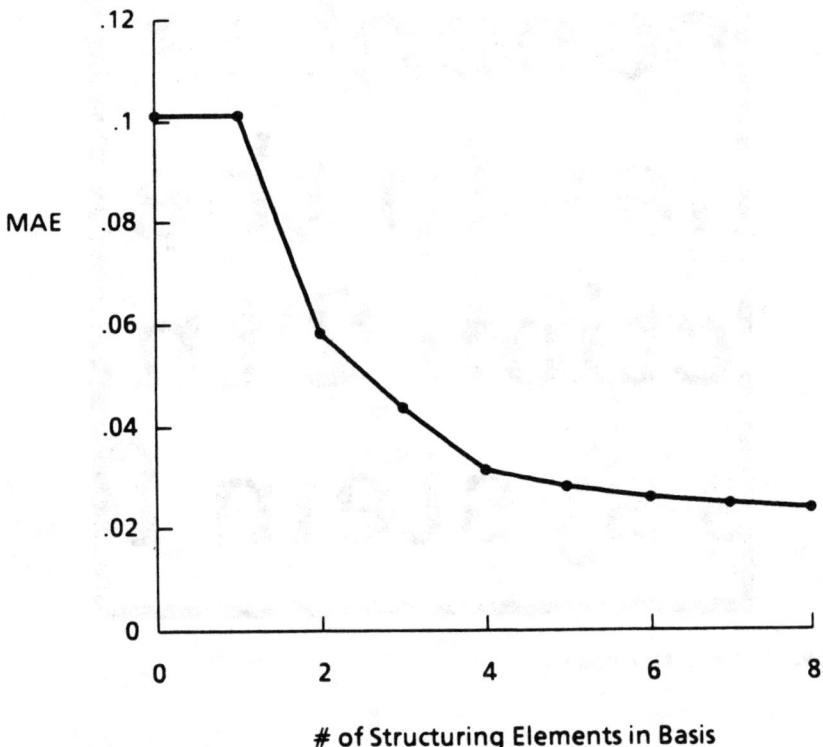

Fig. 10.7. MAE curve.

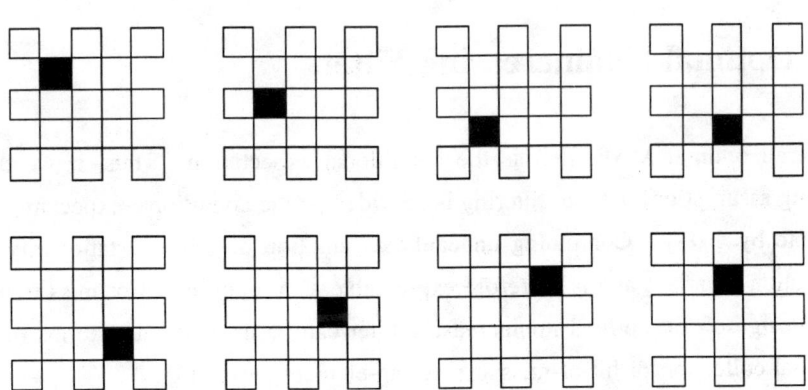

Fig. 10.8. Optimal 8-erosion basis from library.

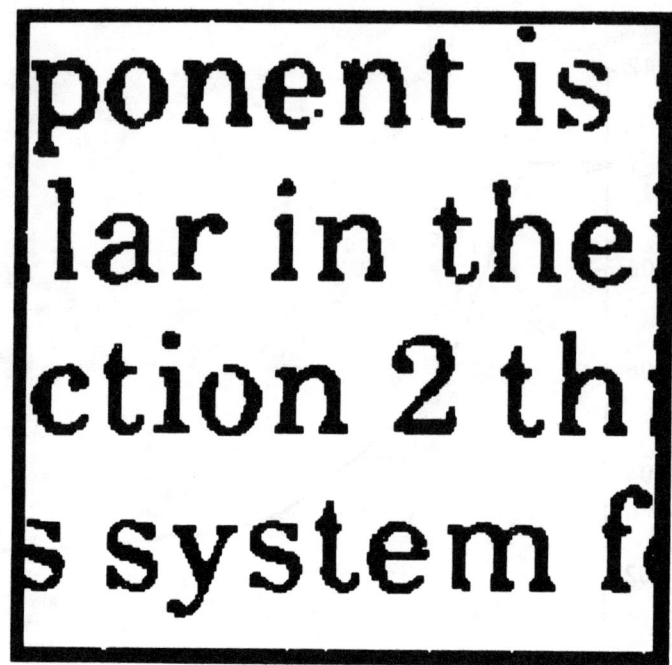

Fig. 10.9. Restored text image.

Although we shall not cover it here, optimal increasing filtering for the gray-scale has also been developed in the framework of the Matheron representation [34] and there exists a gray-scale theorem for MAE representation [110]. Corresponding theories also exist for computational morphological representations [109].

10.5. Optimal Nonincreasing Filters

The overall optimal MAE filter is the conditional expectation. Thus, if we drop the increasing assumption, optimal filtering is provided by the conditional expectation, which we denote by Ψ_{MAE}. Continuing under the assumption of process stationarity, Ψ_{MAE} is translation invariant and is therefore expressable as a union of hit-or-miss transforms. Hence, design of the optimal nonincreasing filter can be understood in terms of finding an optimal collection of hit-or-miss structuring-element pairs [43].

Consider a fixed window W and the collection **C** of all canonical (no don't-care) structuring-element pairs associated with the window. Suppose Ψ is a filter defined as a union of hit-or-miss structuring pairs from $\mathbf{C}$, namely,

$$\Psi(S) = \sum \{ S \circledast (E, F) : (E, F) \in \mathbf{C}_0 \}, \tag{10.15}$$

where $\mathbf{C}_0$ is a subcollection of $\mathbf{C}$. A pixel x lies in $\Psi(S)$, which means numerically that $\Psi(S)(x) = 1$, if and only if there exists a pair (E, F) in $\mathbf{C}_0$ such that $E = S \cap W_x$ and $F = S^c \cap W_x$. Intuitively, this means that $\Psi(S)(x) = 1$ if and only if there exists a pair in the hit-or-miss expansion such that, as a template, the pair matches the windowed portion of the image, both foreground and background.

If we return to Eq. 10.3 and the random-vector description of the observations in the window employed there, then $X_1, X_2, \ldots, X_n$ are the binary observations lying in the translated window, with $X_k = 1$ if and only if the kth pixel in the window lies in the observed image, and Y is the binary value of the ideal image at pixel x, with $Y = 1$ if and only if x lies in the ideal image. According to Eq. 10.3, x lies in the filtered image $\Psi_{\mathrm{MAE}}(S)$ if and only if $P(Y = 1 | X_1, X_2, \ldots, X_n) > 0.5$, the latter meaning that, given the observation $S \cap W_x$, the probability of x lying in the ideal image is greater than 0.5. We can rewrite this numerically as

$$\Psi_{\mathrm{MAE}}(S)(x) = 1 \text{ iff } P(\mathbf{S}_0(x) = 1 \mid S \cap W_x) > 0.5 \tag{10.16}$$

or, equivalently, in terms of random sets as

$$x \in \Psi_{\mathrm{MAE}}(S) \text{ iff } P(x \in \mathbf{S}_0 \mid S \cap W_x) > 0.5, \tag{10.17}$$

where $\mathbf{S}_0$ is the ideal-image process. If we combine this latter statement regarding the conditional-expectation filter with the comments of the preceding paragraph concerning necessary and sufficient conditions for a canonical hit-or-miss expansion to have value 1 (have a pixel lie in the filtered image), we conclude that a structuring pair (E, F) is in the hit-or-miss expansion forming the conditional expectation if and only if

$$P(x \in \mathbf{S}_0 | S \cap W_x = E) > 0.5, \tag{10.18}$$

where $S \cap W_x = E$ means that the observed windowed process equals E and, because the template is canonical, the observed complementary windowed process, $S^c \cap W_x$, equals F.

While notationally cumbersome, the preceding characterization of the hit-or-miss expansion for the conditional expectation is quite straightforward. For the observed windowed region

$$S \cap W_x = \begin{pmatrix} 1 & 1 & 1 \\ 0 & 1 & 1 \\ 0 & 0 & 1 \end{pmatrix} \qquad (10.19)$$

the structuring pair corresponding to the observation is

$$E = \begin{pmatrix} 1 & 1 & 1 \\ 0 & 1 & 1 \\ 0 & 0 & 1 \end{pmatrix} \qquad F = \begin{pmatrix} 0 & 0 & 0 \\ 1 & 0 & 0 \\ 1 & 1 & 0 \end{pmatrix} . \qquad (10.20)$$

If the probability of x lying in the ideal image exceeds 0.5 given the observation $S \cap W_x$ of Eq. 10.19, then the structuring pair (E, F) is part of the hit-or-miss expansion for Ψ_{MAE}; otherwise, it is not.

Since, in the finite case, the hit-or-miss expansion corresponds to a logical sum of products, once the canonical expansion for the conditional expectation has been found, logic reduction can be employed to reduce the expansion (nonuniquely), the result being a hit-or-miss expansion with structuring pairs that are not necessarily canonical.

In practice, the conditional probabilities of Eqs. 10.16 and 10.17 are found from realizations of the ideal process and the observed (noisy) process. An estimate of each conditional probability is given by scanning across observed realizations and recording the fraction of times that the corresponding pixel in the ideal image has value 1 when the template is observed. Filter design is accomplished via Eq. 10.18 using the resulting probability estimates. A key aspect of estimation is that many more realization observations are required in the nonincreasing case than in the increasing case to obtain a satisfactory level of precision in this estimation process [45].

If a degraded image results from union noise, then the optimal filter will be a parallel thinning of the kind given in Eq. 4.17; if the degradation results from subtractive noise, then it will be a parallel thickening of the kind given in Eq. 4.18 [43]. In either case, one needs to apply the optimization procedure to find the structuring pairs forming the corresponding union. The advantage of the thinning or thickening approach (when one knows the degradation is union or subtractive, respectively) is that there will likely be far fewer pairs in the hit-or-miss union. Because the union of Eq. 4.17 gives the pixels to be removed from the observed image, the canonical pair (E, F) is considered for the union only if E contains the origin and it belongs to the union if and only if $P(x \in \mathbf{S}_0 \mid S \cap W_x = E) < 0.5$; on the other hand, because the union of Eq. 4.18 gives the pixels to be adjoined to the observed image, (E, F) is considered for the union only if F contains the origin and it belongs to the union if and only if $P(x \in \mathbf{S}_0 \mid S \cap W_x = E) > 0.5$.

To illustrate the advantage of optimization over human design, we consider two types of edge noise affecting the image of digital "circles" shown in Fig. 10.10. The circles in the ideal realization cover 8% of a 128×128 pixel^2 frame, the radii of the circles being uniformly randomly distributed over $\{5, 6, 7, 8\}$. The union noise process resulting in the degraded realization of Fig. 10.11 randomly adjoins noise pixels at locations that are not connected to other noise pixels and are connected to ideal image pixels, where the connectivity may involve any number of weak neighbors and no more than one strong neighbor. The resulting union noise appears sparse and not "built-up." The optimal parallel thinning over the 3×3 window has been designed for restoration. The result is the set of 20 canonical templates shown in Fig. 10.12. These are logically equivalent to the set of pruners shown in Fig. 4.7 and they produce the restored image of Fig. 10.13. The restoration is excellent, reducing MAE from 3.397% to 7.06×10^{-3}%; in fact, only one noise pixel remains after application of the filter.

Now consider a less restrictive, more realistic edge-noise process. Figure 10.14 shows a noisy realization of the circle process in which the noise has been randomly adjoined to edge pixels with the restrictions concerning isolated noise pixels and noise pixels only being allowed one strongly connected ideal image pixel have been removed. The set of logic-reduced structuring pairs found by optimization for this noise model is shown in Fig. 10.15 and the result of filtering by the corresponding parallel thinning is shown in Fig. 10.16. MAE has been reduced from 3.390% to 1.038%. If instead of employing optimization, we use the traditional pruners, the filtered image of Fig. 10.17 results. It has an MAE of 1.723% and is noisier in appearance than the optimally thinned image of Fig. 10.16. The point is clear: while human design may work for very simple noise models and produce easily "understandable" filters, classical statistically optimized estimation can provide superior filtering in the presence of more complex noise, albeit, perhaps at the cost of less "understandable" filters.

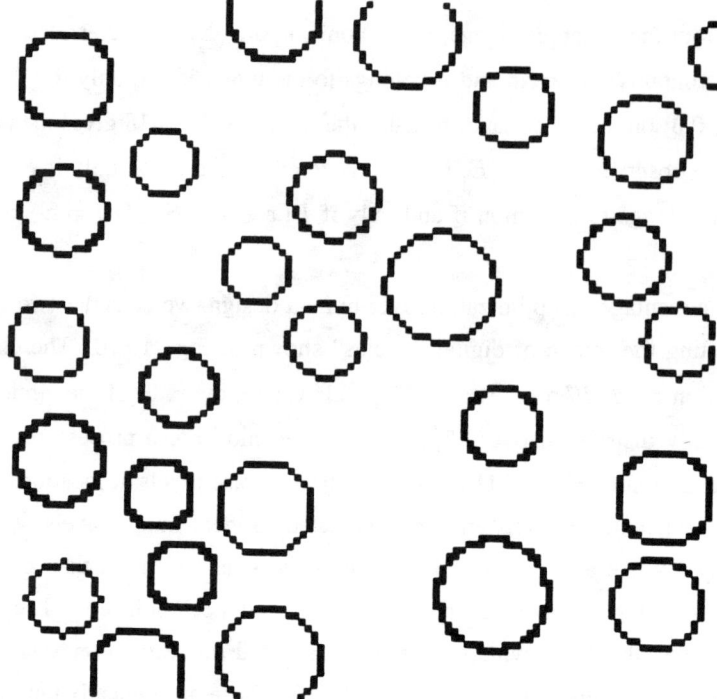

Fig. 10.10. Ideal circle image.

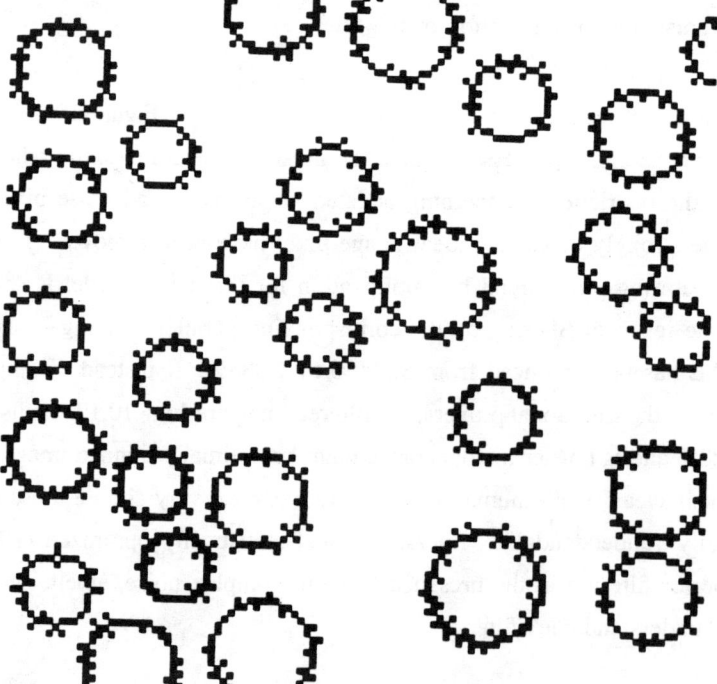

Fig. 10.11. Circle image degraded by restricted edge noise.

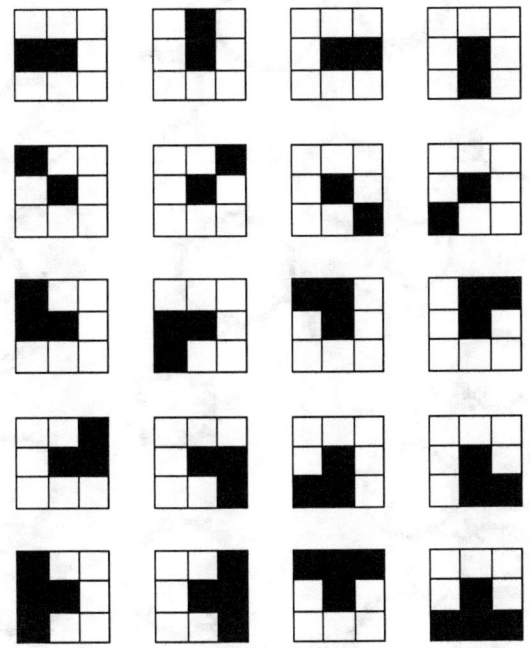

Fig. 10.12. Optimal canonical structuring pairs for restricted noise.

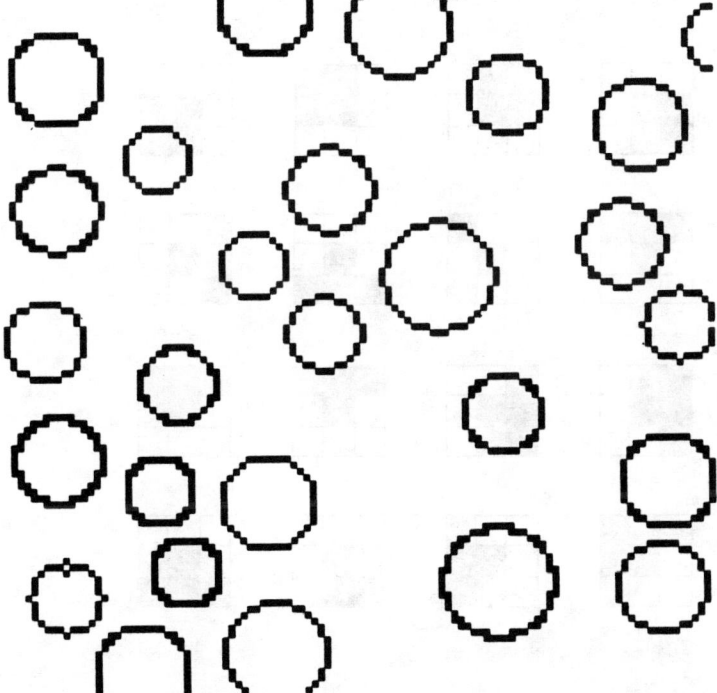

Fig. 10.13. Restored circle image.

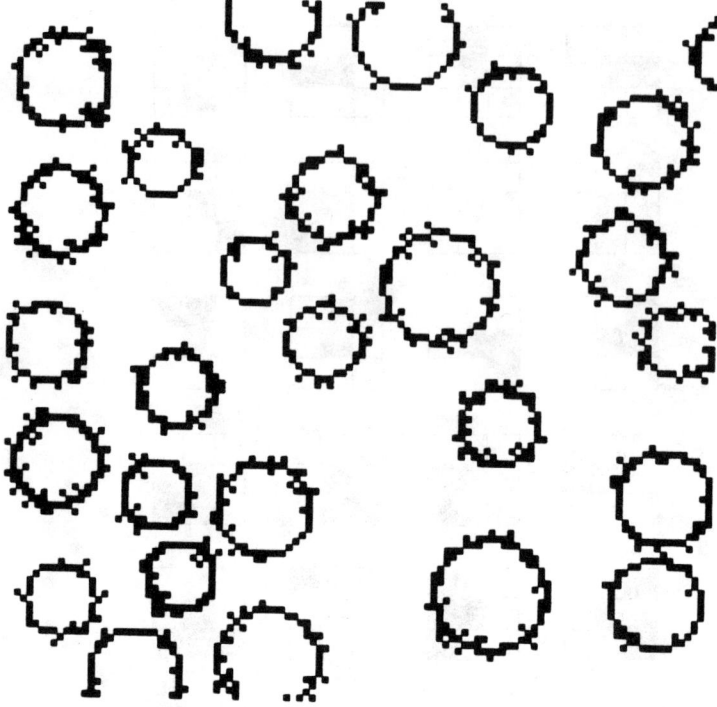

Fig. 10.14. Circle image degraded by unrestricted edge noise.

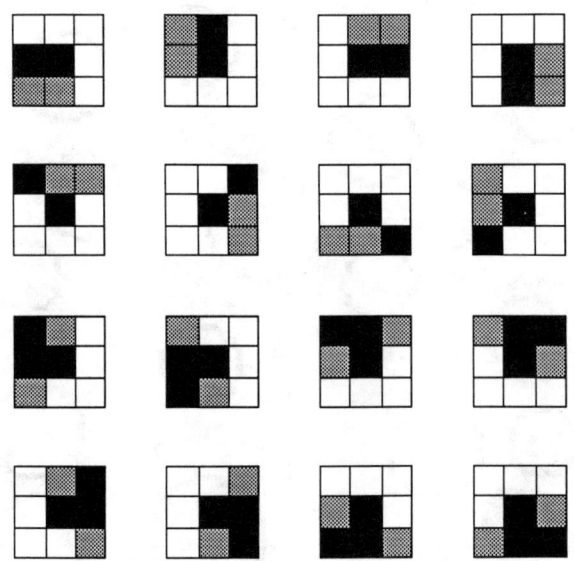

Fig. 10.15. Optimal structuring pairs for unrestricted noise.

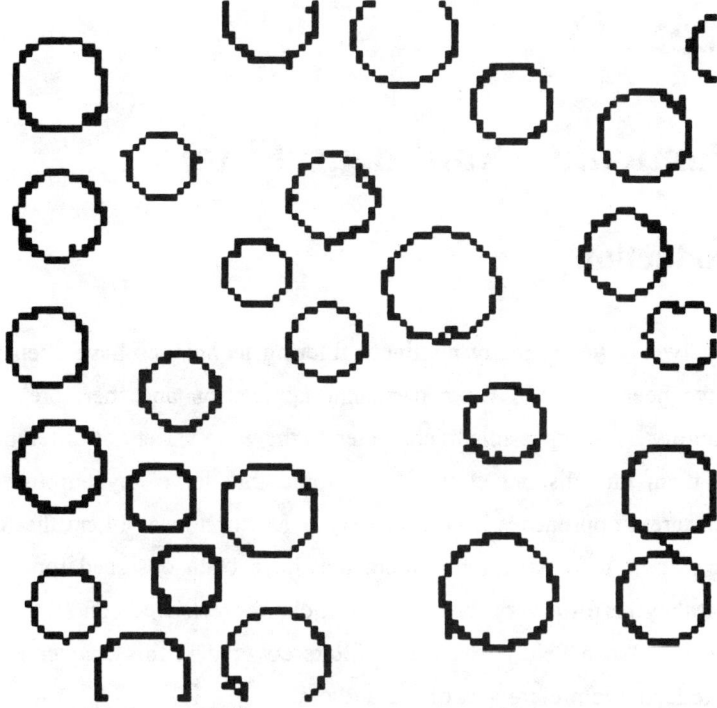

Fig. 10.16. Circle image restored by optimal filter.

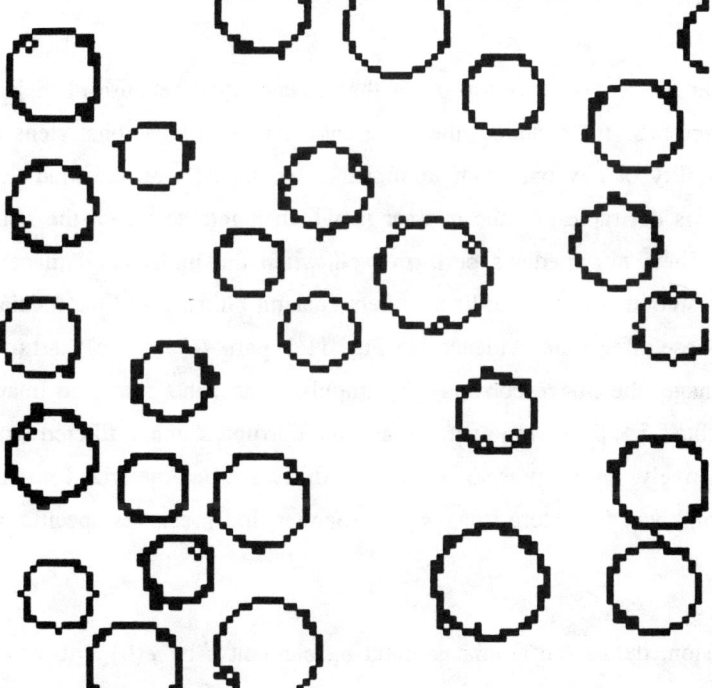

Fig. 10.17. Circle image restored by classical pruners.

Chapter 11

Some Particular Nonlinear Filters

11.1. Introduction

During the last twenty years, many nonlinear filtering techniques have been proposed. Often they have been designed for a particular application and therefore have quite limited applicability. It is impossible to characterize the wide variety of different methods by classifying them into distinct classes, nor would this serve any purpose. Instead, we illustrate different approaches by considering a few nonlinear filters that are useful in image processing. It is typical for filters that have been designed for a particular application that they perform very well under sufficiently restricted conditions but rather poorly outside their "own field." The list of filters covered in this chapter is not at all complete and reflects the preferences of the authors.

11.2. Morphological Pseudoconvolutions

While nonlinear filters often perform better than linear filters because of their structure-preserving properties (for instance, the invariance of one-dimensional steps under the median), the utility of any particular nonlinear filter depends on the kind of structure whose passage is desirable and the manner in which it interacts with the noise. Thus, neither linear filters nor medians perform well when the underlying image is highly textured. Both smooth the texture, linear filters causing blurring and the median causing blotchiness. These effects are evidenced in Fig. 11.1, parts (a), (b), (c), and (d) showing the original image, the image corrupted by impulse noise, the corrupted image filtered by the unweighted 5×5 moving average, and the corrupted image filtered by the 5×5 median, respectively. In the present section we discuss a parameterized nonlinear filter that can be employed to restore images in a manner that preserves specific underlying structure.

In one dimension, define the nonflat structuring element e by $e(0) = 0$ and $e(-1) = e(1) = -\lambda$, where $\lambda \geq 0$ (Fig. 11.2); in two dimensions, define $e(0,0) = 0$ and

$e(i, j) = -\lambda$ for each strong neighbor of the origin. If Ψ is any image operator, define the operators E and D by

$$E(f) = \min(f \ominus e, \Psi(f)) \qquad\qquad (11.1)$$

$$D(f) = \min(f \oplus e, \Psi(f)). \qquad\qquad (11.2)$$

Define the derived filter Ψ' by

$$\Psi'(f) = \min(f \oplus e, E(f)). \qquad\qquad (11.3)$$

Our main concern is when Ψ is a translation-invariant, increasing filter. In particular, if Ψ is the median or mean, then Ψ' is called the **pseudomedian** or **pseudomean**, respectively, or, more generally, Ψ' is termed a **morphological pseudoconvolution** [41]. In the definition of Ψ', E makes small peaks and upward steps invariant and D makes small valleys and downward steps invariant. The invariant class of Ψ' contains the invariant class of Ψ. Specifically, if the pointwise variation of f is bounded by λ or if f is invariant under Ψ, then f is invariant under Ψ'. Many properties of pseudomeans and pseudomedians are known, including their impulse and step responses, their morphological basis representations, a partial converse to the manner in which the invariant class is expanded by the inclusion of signals possessing small variation, and conditions under which the derived filter provides an unbiased estimator [41]. Figure 11.3 shows the corrupted image of Fig. 11.1 (b) filtered by a pseudomedian and pseudomean for a selected value of λ. Notice the better preservation of the original texture.

The definition of Ψ' depends on both e and the choice of λ. Regarding e, one need not be confined to the structuring element defined above. This choice of e has been employed to preserve texture; in fact, in any particular application the shape of e depends on the underlying image structure to be passed.

For filtering scanning tunneling microscope (STM) images by morphological pseudoconvolutions, the shape of e has been based on the atomic structure of graphite, the result being a 5×5 parabolic structuring element [210]. Figures 11.4 (a) - (d) show images of simulated graphite, the graphite corrupted by simulated STM noise, the corrupted image filtered by a Wiener filter, and the corrupted image filtered by a pseudomean, respectively. In each case a cross section is also shown. Notice how the morphologically filtered image contains less scan-line noise and how it has better restored the valleys within the peaks.

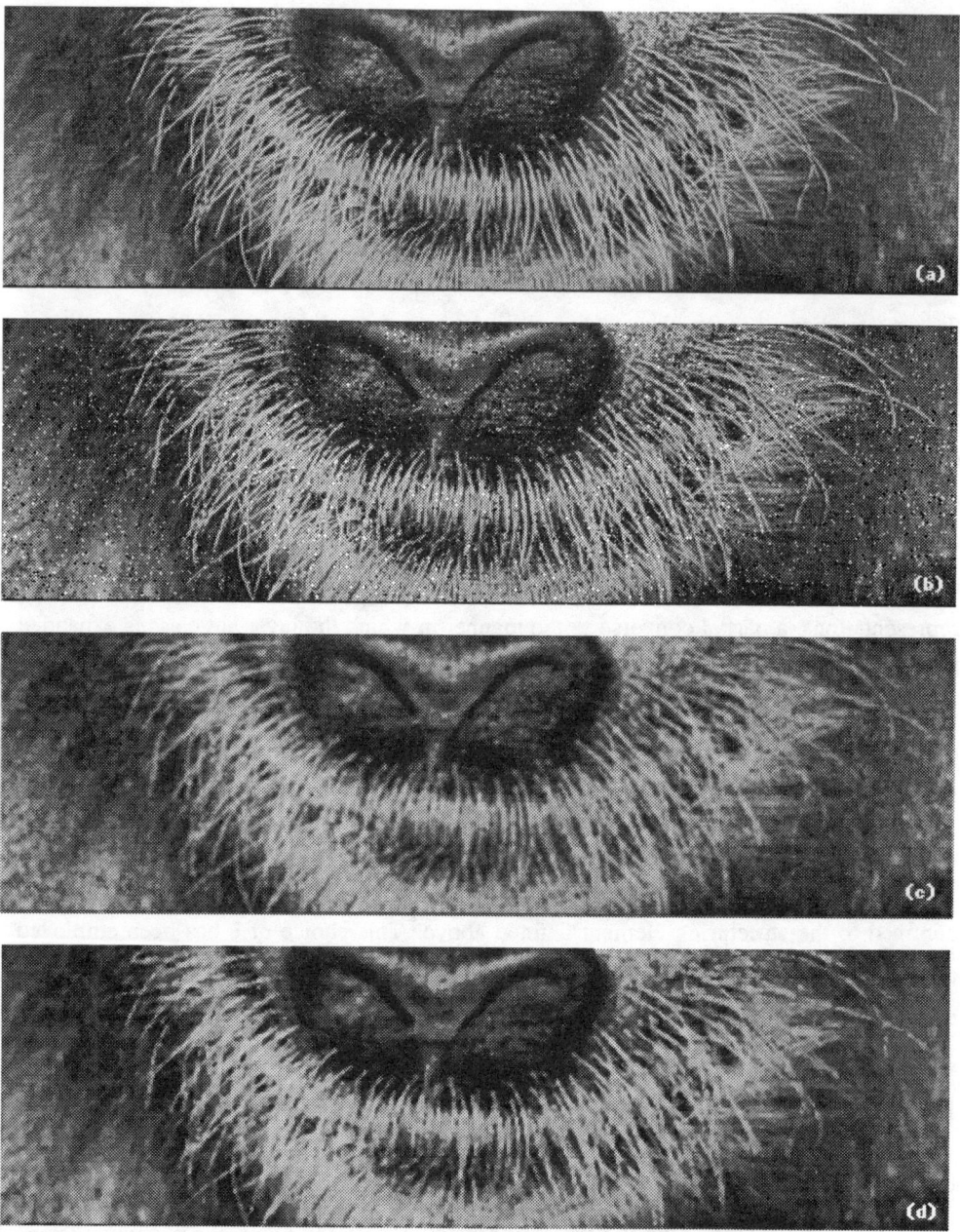

Fig. 11.1. Effects of mean and median on texture degraded by impulse noise: (a) original image, (b) degraded image, (c) mean-filtered image, (d) median-filtered image.

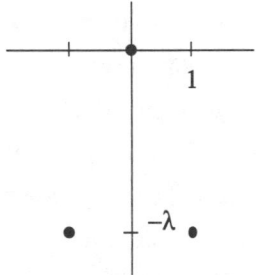

Fig. 11.2. Pseudoconvolution structuring element.

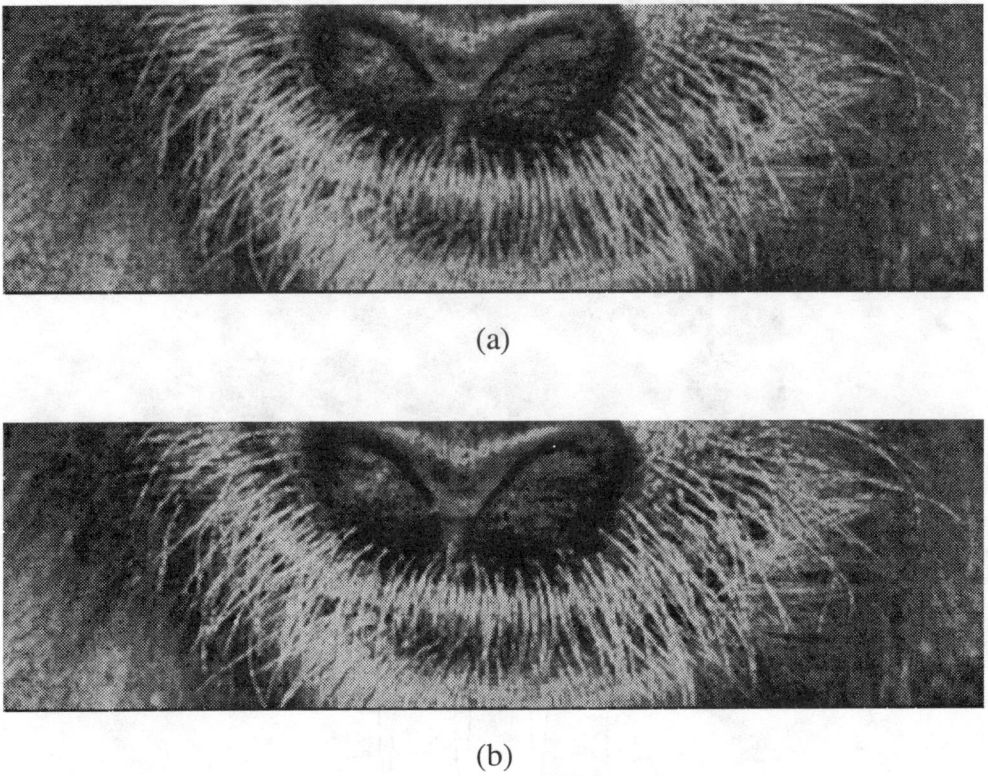

(a)

(b)

Fig. 11.3. Pseudoconvolution filtered images: (a) pseudomedian; (b) pseudomean.

A practical example of pseudomedian filtering of STM images is shown in Fig. 11.5. Part a shows an unfiltered 256×256 image of $3.1 \times 10^{-5} g/mL$ of polyhexylthiophene disolved in toluene. The image shows three strands, one at the left-hand side of the image and two near the center of the image, one apparently crossing the other. In the upper right-hand corner there is some structure appearing as protrusions along the strand but this structure is obscured by noise. Determining the edges of the protrusions is difficult because the presense of scan-line noise causes these edges to appear at different locations in each scan line. Figure 11.5 (b) shows the observed STM image filtered by a pseudomedian. In the filtered image, the edges of the protrusions are fairly uniform from one scan line to the next and one can better quantitatively determine their spacings and relate them to the expected spacing in the polymer.

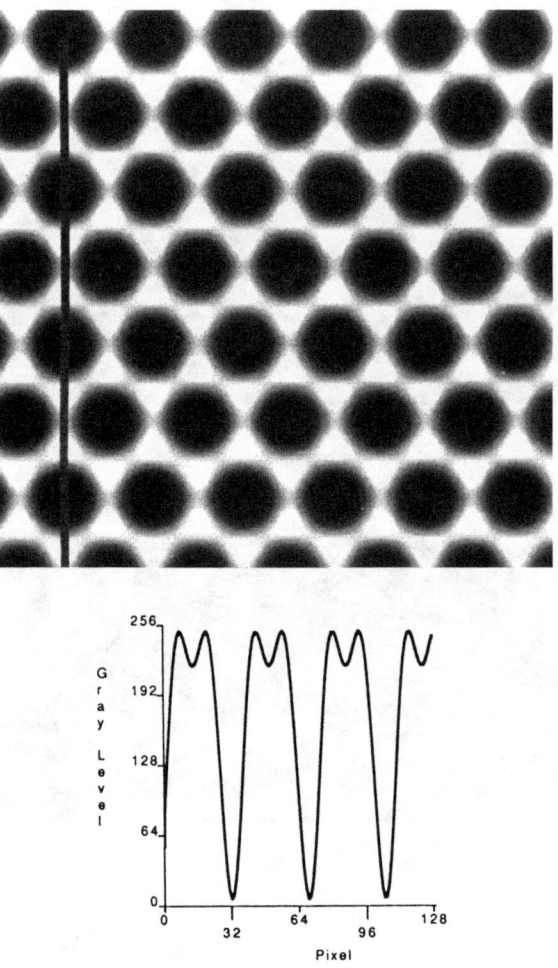

Fig. 11.4. Filtering graphite degraded by STM noise: (a) simulated graphite image.

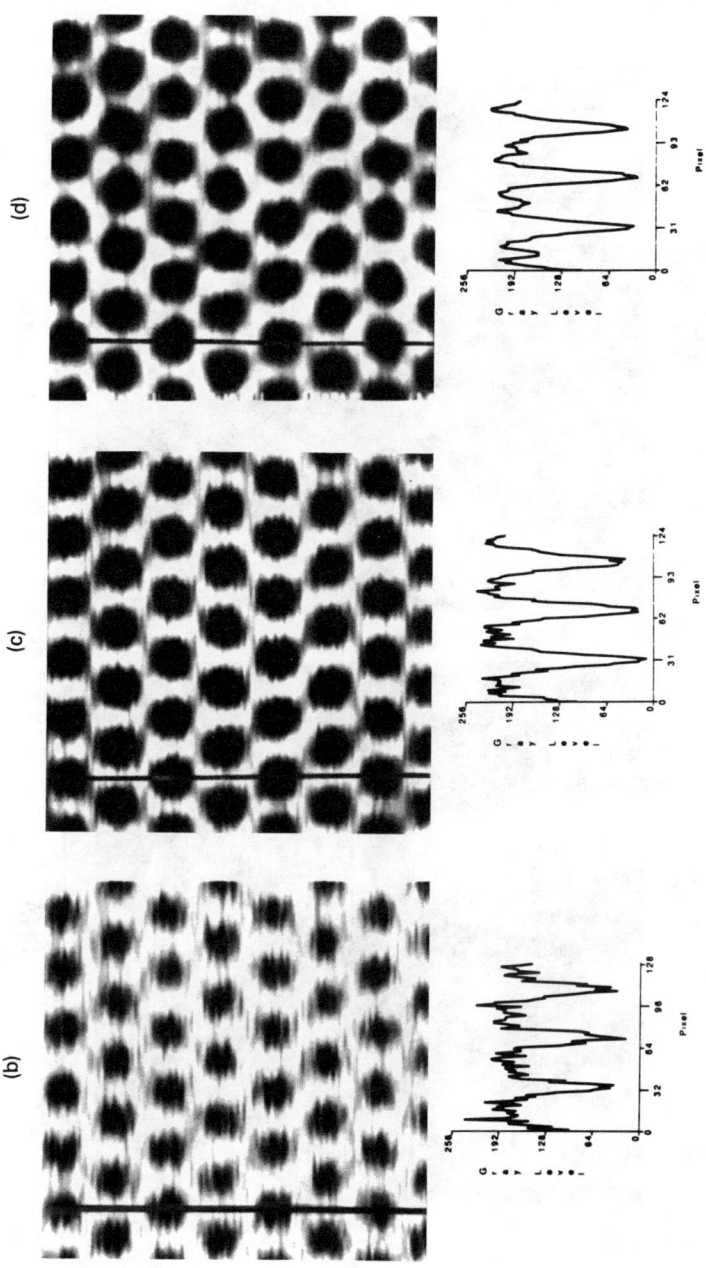

Fig. 11.4. Filtering graphite degraded by STM noise: (b) degraded graphite image; (c) Wiener-filtered image; (d) pseudomean-filtered image.

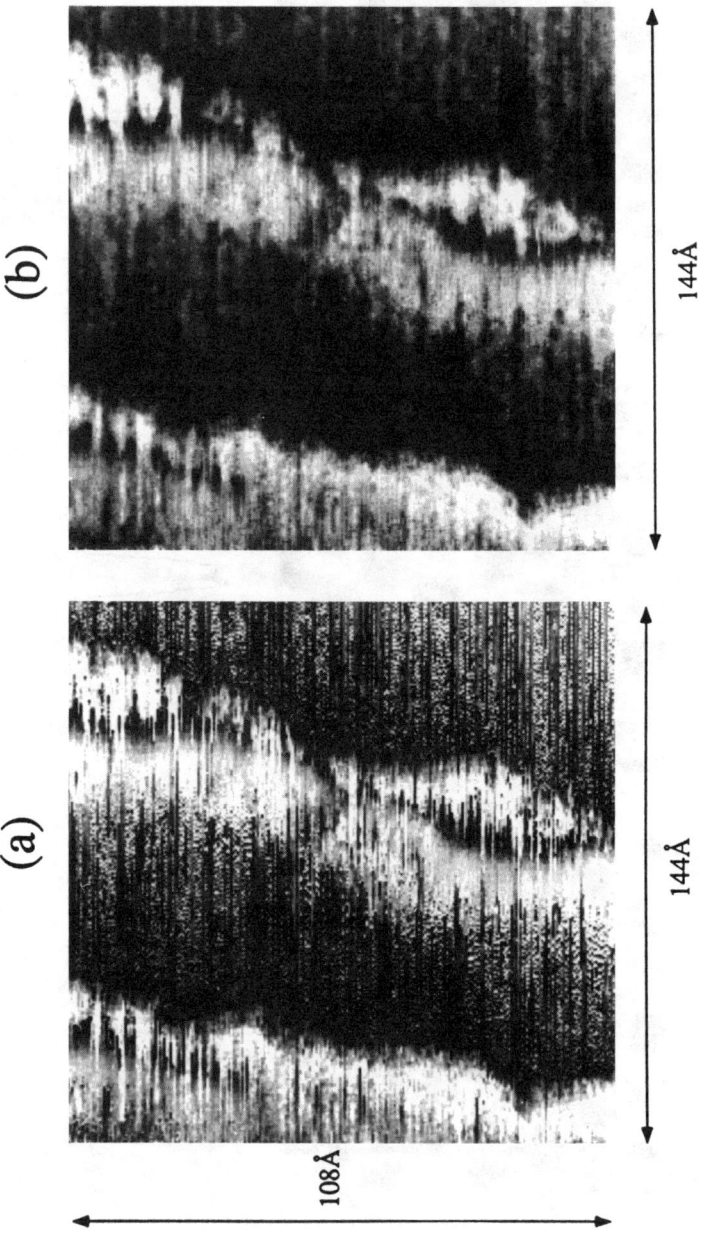

Fig. 11.5. Filtering an STM image of polyhexylthiophene: (a) unfiltered image; (b) filtered image.

Selection of an appropriate value of λ is key to successful application of pseudoconvolutions. This choice will depend on both the noise and the underlying structure one desires to pass. Various families of operating characteristic curves have been developed to facilitate a suitable choice of λ for the structuring element e of Fig. 11.2 [41]. These curves provide optimal λ for various scenarios: suppression of white noise, restoration of edges, and restoration of texture. They are based on image-noise models and must be used with care because often there are competing goals in the restoration process.

11.3. Soft Morphological Filters

Soft morphological filters [3,98,102,103] are morphological operators that behave almost like the basic flat morphological operators - dilation and erosion with a simple structuring element - but are not equally sensitive to noise. They can be used if one has the ideal structuring element for the noise-free situation and wants to remove disturbing noise.

Soft morphological operations are most naturally defined in the framework of weighted-order statistics. The two basic soft morphological operations are soft erosion and soft dilation. The flat structuring element of standard morphology is replaced by the concept of (flat) structuring system, which gives the flexibility required for noise tolerance. The **structuring system** $[B, A, r]$ consists of three parameters, finite sets A and B, $A \subset B$, of $\mathbf{Z} \times \mathbf{Z}$ and a natural number r satisfying $1 \leq r \leq |B|$. The set B is called the structuring set, A its (hard) center, $B \setminus A$ its (soft) boundary, and r the **order index** of its center. Soft morphological operations transform a gray-level image $f : \mathbf{Z} \times \mathbf{Z} \to \mathbf{R}$ to another image.

Soft erosion of f by the structuring system $[B, A, r]$ is denoted by $f \ominus [B, A, r](x)$ and is defined by

$$f \ominus [B, A, r](x) = \text{the } r^{th} \text{ smallest value of the multiset}$$

$$\{r \diamond f(a) : a \in A_x\} \cup \{f(b) : b \in (B \setminus A)_x\}, \text{ for all } x \in \mathbf{Z} \times \mathbf{Z}. \tag{11.4}$$

As in Chapter 6, $\diamond$ denotes the repetition operator and a multiset is a collection of objects where repetition is allowed; e.g., $\{1, 1, 1, 2, 3, 3\} = \{3 \diamond 1, 2, 2 \diamond 3\}$ is a multiset. **Soft**

dilation of f by the structuring system $[B, A, r]$ is denoted by $f \oplus [B, A, r](x)$ and is defined by

$$f \oplus [B, A, r](x) = \text{the } r^{th} \text{ largest value of the multiset}$$
$$\{r \diamond f(a) : a \in A_x\} \cup \{f(b) : b \in (B \setminus A)_x\}, \text{ for all } x \in \mathbf{Z}^2. \tag{11.5}$$

The soft erosion (dilation) of f by the stucturing system $[B, A, r]$ at any point x is obtained by shifting the sets B and A to the location x and forming the multiset from the values of f inside the shifted sets, where the values of f inside the hard center are repeated r times, and then by taking the rth smallest (largest) value of the multiset. Soft erosion and dilation are illustrated in Figs. 11.6 and 11.7 respectively.

In extreme cases, soft morphological operations reduce to standard morphological operations. If $r = 1$ or $A = B$, we have the standard operation by the structuring element B and if $r > |B \setminus A|$, we have the standard operation by the structuring set A.

For a gray-scale example, let the structuring set and its center be

$$B = \{(-1, 0), (0, 1), (0, 0), (0, -1), (1, 0)\},$$
$$A = \{(0, 0)\}. \tag{11.6}$$

Then the soft erosion by the structuring system $[B, A, 4]$ is defined by

$$f \ominus [B, A, 4](x) = \text{the } 4^{th} \text{ smallest value of the multiset}$$
$$\{f(x_1 - 1, x_2), f(x_1, x_2 + 1), f(x_1, x_2), f(x_1, x_2), \tag{11.7}$$
$$f(x_1, x_2), f(x_1, x_2), f(x_1, x_2 - 1), f(x_1 + 1, x_2)\}.$$

The output of this filter at point $x = (x_1, x_2)$ is $f(x)$ unless all the values of the set $\{f(b) : b \in (B - A)_x\}$ are smaller than $f(x)$, in which case the output is the largest value of set $\{f(b) : b \in (B - A)_x\}$.

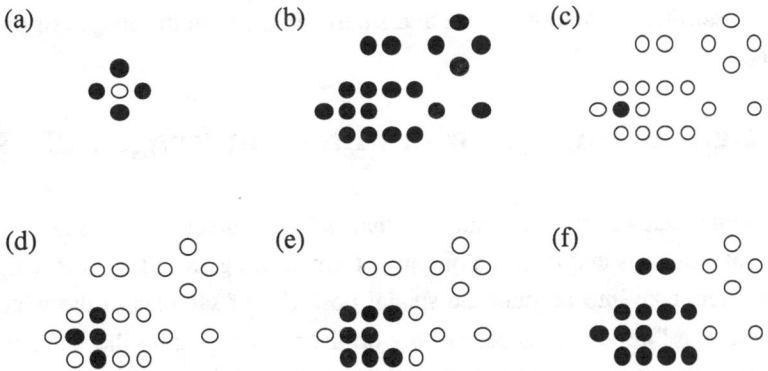

Fig. 11.6. (Provided by Pauli Kuosmanen). Soft erosions of a binary image: (a) the structuring set B and its center A ($\circ$ = an element of the structuring set that marks the origin of $\mathbf{Z}^2$ and the hard center A, $\bullet$ = an element of the structuring set B), (b) the original image ($\bullet$ = an image point), (c) $f \ominus [B, A, 1]$, (d) $f \ominus [B, A, 2]$, (e) $f \ominus [B, A, 3]$, (f) $f \ominus [B, A, 4]$ ($\circ$ = a point that belongs to the original image but not to the erosion).

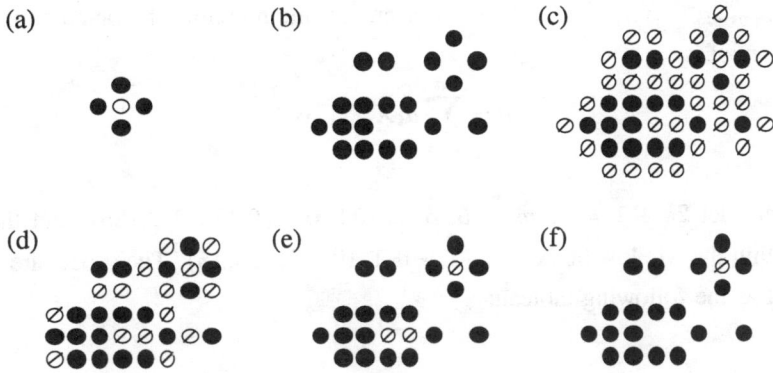

Fig. 11.7. (Provided by Pauli Kuosmanen). Soft dilations of a binary image: (a) the structuring set B and its center A ($\circ$ = an element of the structuring set that marks the origin of $\mathbf{Z}^2$ and the hard center A, $\bullet$ = an element of the structuring set B), (b) the original image ($\bullet$ = an image point), (c) $f \oplus [B, A, 1]$, (d) $f \oplus [B, A, 2]$, (e) $f \oplus [B, A, 3]$, (f) $f \oplus [B, A, 4]$ ($\oslash$ = a point that belongs to the dilation but not to the original image).

As can be seen from the above examples, soft morphological filters behave very similarly to the basic morphological operations. The main difference is that soft morphological filters are less sensitive to additive noise and small variations in the shapes of the objects to be filtered.

11.4. Weighted Majority with Minimum Range Filters

A weighted-order-statistic filter operates so that different pixels within the window are given different emphasis and then the output is formed using an order-statistic operation. Thus, these filters take into account the spatial positions of samples in the window. On the other hand, the "weighing" depends only on the position within the window but not on the relative magnitude of the sample. L-filters give different emphasis to samples solely on the basis of their ranked order. Moreover, the weighing is independent of the relative magnitude of the values within the window.

The intuitive idea of **weighted majority with minimum range (WMMR) filters** [113] is to first find a subset of the values within the window that would best represent the true signal value and then form a linear combination of these values. Let the ordered data values within the window be $X_{(1)}, X_{(2)}, \ldots, X_{(2k+1)}$ and consider $a_1, \ldots, a_m$, where $k+1 \leq m \leq 2k+1$, and such that $a_i \geq 0, i = 1, \ldots m$ and $\sum a_i = 1$. First calculate the ranges $X_{(i+m-1)} - X_{(i)}$, $i = 1, \ldots 2k - m$ and let the minimum be obtained for $i = I$. Then the output is

$$y = \sum_{i=1}^{m} a_i X_{(I+i-1)}.$$

For example, let $2k + 1 = 9$, $m = 5$, $\mathbf{a} = (0.5, 0.25, 0.125, 0, 0.125)$, and the signal section within the window be $\mathbf{X} = (-1, 9, 0, 1, 10, -8, 1, 2, 15)$. The procedure is easily carried out as the following tableau:

i	1	2	3	4	5	6	7	8	9
X_i	-1	9	0	1	10	-8	1	2	15
$X_{(i)}$	-8	-1	0	1	1	2	9	10	15
$X_{(i+5)} - X_{(i)}$	9	3	9	9	14				

minimum range	-1	0	1	1	2
a_i	0.5	0.25	0.125	0.0	0.125

Output
$$0.5(-1) + 0.25(0) + 0.125(1) + \\ + 0.0(1) + 0.125(2) = -0.125$$

The WMMR filter is closely related to the least median of squares method in estimation theory. Among interesting and useful properties of these filters are the facts that they tolerate outliers very well, their root signal sets consist of piecewise constant signals, and, moreover, repeated filtering produces piecewise constant signals. On the other hand, both L-filters and WMMR filters contain additive operations, which means that unlike morphological or stack filters they are not invariant to monotonic transformations. In image processing this means that histogram transformations affect the way the filters operate.

11.5. L-Filters

Let $X_1, X_2, \ldots, X_N$ be the signal values in the window and $X_{(1)}, X_{(2)}, \ldots, X_{(N)}$ be the same values in nondecreasing order. The output of the L-filter of window size $N = 2k+1$ and coefficients a_i, $i = 1, \ldots, N$ is defined by

$$y = \sum_{i=1}^{N} a_i X_{(i)}. \tag{11.8}$$

It is a generalization of the median or any other ranked-order operation, as well as the midpoint or α-trimmed mean, as is seen by considering the following choices:

$$a_i = \begin{cases} 1 & \text{if } i = k+1 \\ 0 & \text{otherwise} \end{cases} \quad \text{(median)}$$

$$a_i = \begin{cases} 1 & \text{if } i = r,\, 1 \leq r \leq N \\ 0 & \text{otherwise} \end{cases} \quad \text{(r-th ranked order)}$$

$$a_i = \begin{cases} \frac{1}{2} & \text{if } i = 1 \text{ or } N \\ 0 & \text{otherwise} \end{cases} \quad \text{(midpoint)}$$

and

$$a_i = \begin{cases} \frac{1}{n(1-2\alpha)} & i = \alpha n + 1, \ldots, n - \alpha n \\ 0 & \text{otherwise} \end{cases} \quad \text{(α-trimmed mean)}$$

where the choice $\alpha = 0$ gives the simple average.

The statistical properties of L-filters have been quite extensively studied [cf. 143 and the references thereof]. For particular noise distributions L-filters perform slightly better than the standard median filter but the difference is rather small.

11.6. Linear-Median Hybrid Filters

When the window size is large, the filters described in previous sections involve extensive computations and often become too slow for real-time applications. A way to solve this problem is to combine fast filters (in particular linear filters) with order-statistic filters in such a way that the combined operation is reasonably close to the desired operation but is significantly faster.

In linear-median hybrid filtering [2,84,137] a cascade of linear and median filters is used so that a small number of linear filters operate over large windows and the median of the outputs of the linear filters is the output of the linear-median hybrid filter.

Consider a one-dimensional signal $x(n)$. The linear-median hybrid filter with substructures $\Phi_1, \ldots, \Phi_M$ is defined by

$$y(n) = \text{MED} [\Phi_1(x(n)), \Phi_2(x(n)), \ldots, \Phi_M(x(n))], \qquad (11.9)$$

where $\Phi_1, \ldots, \Phi_M$ (M odd) are linear filters. The subfilters Φ_i are chosen so that an acceptable compromise between noise reduction and root signal set is achieved while keeping M small enough to allow simple implementation. As an example, let us look at the following structure:

$$y(n) = \text{MED} [\Phi_L(x(n)), \Phi_C(x(n)), \Phi_R(x(n))].$$

The filters Φ_L, Φ_R are low-pass filters following slower trends of the input signal. The filter Φ_C is designed to react fast to signal level changes, allowing the whole filter to move swifty from one level to another. The subscripts L, C, and R denote left, center and right, indicating the corresponding filter position with respect to the current output value, as shown schematically in Fig.11.8.

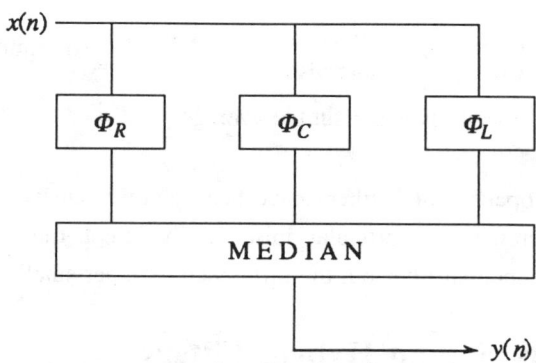

Fig. 11.8. Computing schema of the basic linear-median hybrid filter with subfilters Φ_L, Φ_R, and Φ_C.

The simplest structure introduced in Ref. [84] consists of identical averaging filters Φ_L and Φ_R with $\Phi_C(x(n)) = x(n)$. In time domain the filter is characterized by

$$y(n) = \text{MED} \left[\frac{1}{k} \sum_{i=1}^{k} x(n-k), x(n), \frac{1}{k} \sum_{i=1}^{k} x(n+k) \right]. \tag{11.10}$$

The behavior of this filter is remarkably similar to the behavior of the standard median filter of length $2k+1$ but is very fast to compute. In fact, using recursive running sums, the complexity is constant and thus independent of the window size.

In image processing the number of subfilters is usually five. For example the filter

$$y(n) = \text{MED} \left[\frac{1}{2}(x(m,n-2) + x(m,n-1)), \frac{1}{2}(x(m,n+1) + x(m,n+2)), x(n), \right.$$

$$\left. \frac{1}{2}(x(m+2,n) + x(m+1,n)), \frac{1}{2}(x(m-1,n) + x(m-2,n)) \right]. \tag{11.11}$$

corresponds to the mask shown in Fig. 11.9. Other typical mask shapes are shown in Fig. 11.10.

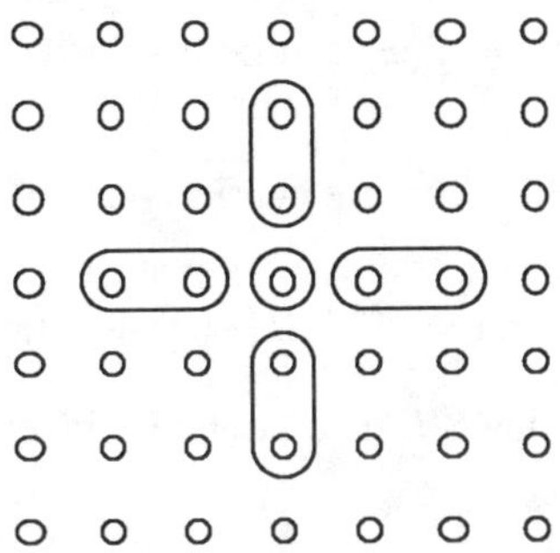

Fig. 11.9. The mask corresponding to the linear-median hybrid filter of Eq. 11.11.

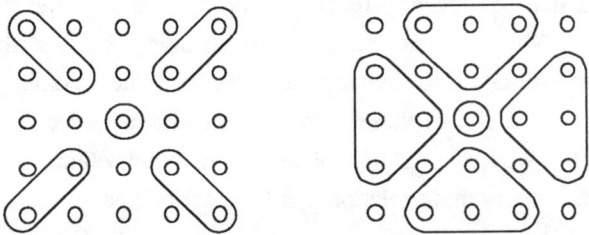

Fig. 11.10. Typical mask shapes of linear-median hybrid filters.

11.7. Homomorphic Filters

Linear filtering techniques are very efficient for additive Gaussian noise, though in image processing their tendency to blur edges is a limiting factor. Often noise is combined with the image in a nonlinear way. A typical example is source illumination, which with object reflectance contributes to image formation in a multiplicative way. In homomorphic filtering [cf. 143], a logarithmic nonlinearity is first applied to the image to transform the multiplicative noise into additive noise. Noise is attenuated by a linear filter and the inverse nonlinear transform is used to get the original "noise-free" image.

Consider the model where the observed image $r(x)$ can be expressed as

$$r(x) = f(x)[1 + n(x)], \tag{11.12}$$

where $f(x)$ is the true image and the noise process $n(x)$ is assumed to satisfy $|n(x)| << 1$. Taking logarithms on both sides gives

$$\log r(x) = \log f(x) + \log [1 + n(x)] \approx \log f(x) + n(x). \tag{11.13}$$

If the noise process $n(x)$ can be successfully eliminated from $\log r(x)$, it is clear that a good approximation of the image $f(x)$ can be obtained.

The principle of homomorphic filtering works whenever the noise formation can be modeled as

$$r(x) = H^{-1}[H(f(x) + N(x)], \tag{11.14}$$

where $r(x)$ is the observed image, H is a nonlinear invertible transformation, and $N(x)$ is the corrupting noise process.

11.8. Polynomial Filters

A class of filters that has recently received considerable attention in image processing is Volterra filters. Volterra filters can, loosely speaking, be described as follows. In linear filtering we estimate a particular pixel value as a linear combination of a fixed set (perhaps all) of observed pixel values. In Volterra filtering we estimate the pixel value with a multivariable series expansion of the set of pixel values. For one-dimensional signals the general input-output relation can be described as

$$y(n) = h_0 + \sum_{k=1}^{\infty} \overline{h}_k[x(n)], \tag{11.15}$$

where

$$\overline{h}_k[x(n)] = \sum_{i_1=0}^{\infty} \cdots \sum_{i_k=0}^{\infty} h_k(i_1, \ldots, i_k) x(n - i_1) \cdots x(n - i_k).$$

In practice the series is truncated and the input x is assumed to have finite support, whence the expression reduces to a polynomial. Note that the term $\overline{h}_1[x(n)]$ in Eq 11.15 corresponds to the impulse response of a linear filter. Similarly, the third term describes the quadratic relation of input samples contributing to the output.

Truncated Volterra filters are called **polynomial filters** [cf. 143] and for low order they are simple to implement. For example, a quadratic polynomial filter has input-output relation

$$y(n) = \sum_{i=0}^{N-1} \sum_{j=0}^{N-1} h_{ij} x(n - i) x(n - j).$$

Because quadratic nonlinearities occur naturally in many imaging systems, it is plausible that polynomial filters will have a role as part of image processing systems.

Bibliography

[1] Astola, J., and Campbell, T. G., "On Computation of the Running Median," *IEEE Trans. Acoust., Speech, Signal Processing*, Vol. ASSP-37, pp. 572-574, Apr. 1989.

[2] Astola, J., Heinonen, P., and Neuvo, Y., "Linear Median Hybrid Filters," *IEEE Trans. Circuits and Systems*, Vol. CAS-36, pp. 1430-1438, Nov. 1989.

[3] Astola, J., Koskinen, L., and Neuvo, Y., "Statistical Properties of Discrete Morphological Filters," in *Mathematical Morphology in Image Processing*, ed. E. R. Dougherty, Marcel Dekker, New York, 1993.

[4] Astola, J., and Neuvo, Y., "Optimal Median Type Filters for Exponential Noise Distributions," *Signal Processing*, Vol. 17, pp. 95-104, 1989.

[5] Astola, J., and Neuvo, Y., "An Efficient Tool for Analyzing Weighted Median and Stack Filters," *IEEE Trans. Circuits and Systems*, submitted.

[6] Ataman, E., Aatre, V. K., and Wong, K. M., "A Fast Method for Real-Time Median Filtering," *IEEE Trans. Acoust., Speech, Signal Processing*, Vol. ASSP-28, pp. 415-421, Aug. 1980.

[7] Bapeswara Rao, V. V., and Sankara Rao, K., "A New Algorithm for Real-Time Median Filtering," *IEEE Trans. Acoust., Speech, Signal Processing*, Vol. ASSP-34, pp. 1674-1675, Dec. 1986.

[8] Banon, G. J. F., and Barrera, J., "Decomposition of Mappings Between Complete Lattices by Mathematical Morphology, Part I. General Lattices," *Journal of Signal Processing*, Vol. 30, February, 1993.

[9] Banon, G. J. F., and Barrera, J., "Minimal Representations of Translation-Invariant Set Mappings by Mathematical Morphology," *SIAM Journal on Applied Mathematics*, Vol. 51, No. 6, December, 1991.

[10] Barrera, J., and Banon, G. J. F., "Expressiveness of the Morphological Language," *Proc. SPIE*, Vol. 1769, pp. 264-275, July, 1992.

[11] Bettoli, B., and Dougherty, E. R., "Linear Granulometric Moments of Noisy Binary Images," *Journal of Mathematical Imaging and Vision*, Vol. 3, No. 3, 1993.

[12] Bhagvati, C., Grivas, A., and Skolnick, M., "Morphological Analysis of Pavement Surface Condition," in *Mathematical Morphology in Image Processing*, ed. E. R. Dougherty, Marcel Dekker, New York, 1993.

[13] Bloch, I., "Triangular Norms as a Tool for Constructing Fuzzy Mathematical Morphologies," *Proc. EURASIP Workshop on Mathematical Morphology and its Applications to Signal Processing*, Barcelona, May, 1993.

[14] Bloch, I., and Maitre, H., "Mathematical Morphology on Fuzzy Sets," *Proc. EURASIP Workshop on Mathematical Morphology and its Applications to Signal Processing*, Barcelona, May, 1993.

[15] Bovik, A. C., "Streaking in Median Filtered Images," *IEEE Trans. Acoust., Speech, Signal Processing*, Vol. ASSP-35, pp. 493-503, Apr. 1987.

[16] Bovik, A. C., Huang, T. S., and Munson, D. C., Jr., "A Generalization of Median Filtering Using Linear Combinations of Order Statistics," *IEEE Trans. Acoust., Speech, Signal Processing*, Vol. ASSP-31, pp. 1342-1350, Dec. 1983.

[17] Brandt, J., "Invariant Signals for Median Filters," *Utilitas Mathematica*, Vol. 31, pp. 93-105, 1987.

[18] Brownrigg, D. R. K., "The Weighted Median Filter," *Communications of the ACM*, Vol. 27, pp. 807-818, Aug. 1984.

[19] Chen, Y., Dougherty, E. R., Totterman, S., and Hornak, J., "Classification of Trabecular Structure in Magnetic Resonance Images Based on Morphological Granulometries," *Journal of Magnetic Resonance Medicine*, Vol. 29, No. 3, March, 1993.

[20] Chen, Y., and Dougherty, E. R., "Texture Classification by Gray-Scale Morphological Granulometries," *Proc. SPIE Visual Communication and Image Processing*, SPIE Vol. 1818, pp. 931-943, November, 1992.

[21] Costa, W. S., and Haralick, R. M., "Predicting Expected Gray Level Statistics of Opened Signals," *Proc. IEEE Conference on Computer Vision and Pattern Recognition*, Champaign, June, 1992.

[22] Crimmons, T., and Brown, W., "Image Algebra and Automatic Shape Recognition," *IEEE Transactions on Aerospace and Electronic Systems*, Vol. 21, January, 1985.

[23] Danielsson, P.-E., "Getting the Median Faster," *Computer Graphics and Image Processing*, Vol. 17, no. 1, pp. 71-78, Sept. 1981.

[24] David, H. A., *Order Statistics*, New York, Wiley, 1978.

[25] Davidson, J., "Morphology Neural Networks: An Introduction with Applications," *Journal of Circuits, Systems, and Signal Processing*, Vol. 12, No. 2, 1993.

[26] Dorst, L., and van den Boomgaard, R., "An Analytic Theory of Mathematical Morphology," *Proc. EURASIP Conference on Mathematical Morphology and its Applications to Signal Processing*, Barcelona, May, 1993.

[27] Dougherty, E. R., editor, *Mathematical Morphology in Image Processing*, Marcel Dekker, New York, 1993.

[28] Dougherty, E. R., *An Introduction to Morphological Image Processing*, SPIE Press, Bellingham WA, 1992.

[29] Dougherty, E. R., "Solution of Morphological Operator Relations with Invariance Boundary Conditions," *Proc. SPIE*, Vol. 2030, pp. 33-40, July, 1993.

[30] Dougherty, E. R., "Optimal Morphological Hit-or-Miss Filtering of Gray-Scale Images," *Proc. SPIE*, Vol. 1902, pp. 30-40, February, 1993.

[31] Dougherty, E. R., "Unification of Nonlinear Filtering in the Context of Binary Logic Calculus – Part II: Gray-Scale Filters," *Journal of Mathematical Imaging and Vision*, Vol. 2, No. 2, December, 1992.

[32] Dougherty, E. R., "Euclidean Gray-Scale Granulometries: Representation and Umbra Inducement," *Journal of Mathematical Imaging and Vision*, Vol. 1., No. 1, March, 1992.

[33] Dougherty, E. R., "Optimal Mean-Square N-Observation Digital Morphological Filters – Part I: Optimal Binary Filters," *Journal of Computer Vision, Graphics, and Image Processing - Image Understanding*, Vol. 55, No. 1, January, 1992.

[34] Dougherty, E. R., "Optimal Mean-Square N-Observation Digital Morphological Filters – Part II: Optimal Gray-Scale Filters," *Journal of Computer Vision, Graphics, and Image Processing - Image Understanding*, Vol. 55, No. 1, January, 1992.

[35] Dougherty, E. R., and J. Astola, editors, *Mathematical Nonlinear Image Processing*, Kluwer Academic Publishers, 1993.

[36] Dougherty, E., and Giardina, C., *Image Processing - Continuous to Discrete*, Prentice-Hall, Englewood Cliffs, 1987.

[37] Dougherty, E., and Giardina, C., "A Digital Version of the Matheron Representation Theorem for Increasing Tau-Mappings in Terms of a Basis for the Kernel," *Proc. IEEE Conference on Computer Vision and Pattern Recognition*, 1986.

[38] Dougherty, E.R., and Haralick, R.M., "Hole-Spectrum Representation and Model-Based Optimal Morphological Restoration for Binary Images Degraded By Subtractive Noise," *Journal of Mathematical Imaging and Vision*, Vol. 1, No. 3, August, 1992.

[39] Dougherty, E. R., and Haralick, R. M., "Unification of Nonlinear Filtering in the Context of Binary Logical Calculus – Part I: Binary Filters," *Journal of Mathematical Imaging and Vision*, Vol. 2, No. 2, December, 1992.

[40] Dougherty, E. R., Haralick, R. M., Chen, Y., Agerskov, C., Jacobi, U., and P. H. Sloth, "Estimation of Optimal Tau-Opening Parameters Based on Independent Observation of Signal and Noise Pattern Spectra," *Journal of Signal Processing*, Vol. 29, December, 1992.

[41] Dougherty, E. R. and Kraus, E., "Morphological Pseudoconvolutions: One-Parameter Families of Derived Filters With Increased Invariant Classes," *Journal of Circuits, Systems, and Signal Processing*, Vol. 11, No. 1, February, 1992.

[42] Dougherty, E. R., and Kraus, E., "Shape Analysis and Reduction of the Morphological Basis for Digital Moving Averages," *SIAM Journal On Applied Mathematics*, Vol. 51, No. 6, December, 1991.

[43] Dougherty, E. R., and Loce, R. P., "Optimal Mean-Absolute-Error Hit-or-Miss Filters: Morphological Representation and Estimation of the Binary Conditional Expectation," *Journal of Optical Engineering*, Vol. 32, No. 4, April, 1993.

[44] Dougherty, E. R., and Loce, R. P., "Robustness of Optimal Binary Morphological Filters Relative to Basis Size," *Twenty Seventh Annual Conference On Information Sciences and Systems*, Baltimore, March, 1993.

[45] Dougherty, E. R., and Loce, R. P., "Precision of Morphological Estimation," *Proc. SPIE*, Vol. 1902, pp. 65-76, February, 1993.

[46] Dougherty, E. R., and Loce, R. P., "Efficient Design Strategies for the Optimal Binary Digital Morphological Filter: Probabilities, Constraints, and Structuring-Element Libraries," *Mathematical Morphology in Image Processing*, ed. E. R. Dougherty, Marcel Dekker, New York, 1993.

[47] Dougherty, E. R. and Loce, R. P., "Robust Morphologically Continuous Fourier Descriptors Generated by Geometric Projections – Part I: The Descriptors and Point Noise Analysis," *International Journal of Pattern Recognition and Artificial Intelligence*, Vol. 6, No. 5, December, 1992.

[48] Dougherty, E. R. and Loce, R. P., "Robust Morphologically Continuous Fourier Descriptors Generated by Geometric Projections – Part II: Hausdorff-Metric Continuity and Occlusion Analysis," *International Journal of Pattern Recognition and Artificial Intelligence*, Vol. 6, No. 5, December, 1992.

[49] Dougherty, E., Mathew, A., and Swarnakar, V., "A Conditional- Expectation-Based Implementation of the Optimal Mean-Square Binary Morphological Filter," *Proc. SPIE*, Vol. 1451, pp. 137-145, February, 1991.

[50] Dougherty, E. R., Newell, J., and Pelz, J., "Morphological Texture-Based Maximum-Likelihood Pixel Classification Based on Local Granulometric Moments," *Journal of Pattern Recognition*, Vol. 25, No. 10, November, 1992.

[51] Dougherty, E. R., Pelz, J., Sand, F., and Lent, A., "Morphological Image Segmentation by Local Granulometric Size Distributions," *Journal of Electronic Imaging*, Vol. 1. No. 1, January, 1992.

[52] Dougherty, E., and Pelz, J., "Morphological Granulometric Analysis of Electrophotographic Images – Size Distribution Statistics For Process Control," *Optical Engineering*, Vol. 30, No. 4, April, 1991.

[53] Dougherty, E. R., and Sand, F., "Moment Representation for Linear Granulometries," *Twenty Seventh Annual Conference On Information Sciences and Systems*, Baltimore, March, 1993.

[54] Dougherty, E. R., and Sinha, D., "Computational Morphology: A Unified Approach to Filter Design," *Proc. EURASIP Workshop on Mathematical Morphology and its Application to Signal Processing*, Barcelona, May, 1993.

[55] Dougherty, E. R., and Sinha, D., "Representation of Finite-Range Increasing Filters in the Context of Computational Morphology," *Proc. SPIE*, Vol. 1902, pp. 53-64, February, 1993.

[56] Dougherty, E. R., and Sinha, D., "Computational Mathematical Morphology," *Morphological Imaging Laboratory Report*, No. MIL-17-92, Rochester Institute of Technology, Rochester, December, 1992.

[57] Dougherty, E. R., Sinha, D., and Sinha, P., "Fuzzy Morphological Filters," *Proc. SPIE Intelligent Robots and Computer Vision: Algorithms and Techniques*, Proc. SPIE, Vol. 1825, pp. 414-426, November, 1992.

[58] Dougherty, E. R., and Zhao, D., "Model-Based Characterization of Statistically Optimal Design for Morphological Shape Recognition Via the Hit-or-Miss Transform," *Journal of Visual Communication and Image Representation*, Vol. 3, No. 2, June, 1992.

[59] Eberly, D., Longbotham, H., and Aragon, J., "Complete Classification of Roots to One-Dimensional Median and Rank-Order Filters," *IEEE Trans. Signal Processing*, Vol. 39, pp. 197-200, Jan. 1991.

[60] Fitch, J. P., Coyle, E. J., and Gallagher, N. C., "Root Properties and Convergence Rates for Median Filters," *IEEE Trans. Acoust., Speech, Signal Processing*, Vol. 33, pp. 230-240, Feb. 1985.

[61] Gader, P., "Template Generation for Pattern Classification," *Proc. SPIE*, Vol. 1769, pp. 72-81, July, 1992.

[62] Gallagher, N. C., and Wise, G. L., "A Theoretical Analysis of the Properties of the Median Filter," *IEEE Trans. Acoust., Speech, Signal Processing*, Vol. 29, pp. 1135-1141, Dec. 1981.

[63] Giardina, C., and Dougherty, E., *Morphological Methods in Image and Signal Processing*, Prentice-Hall, Englewood Cliffs, 1988.

[64] Gil, J., and Werman, M., "Computing 2-D Min, Median and Max Filters, *IEEE Trans. Pattern Anal. and Machine Intellig.* Vol. 15, No. 5, May 1993.

[65] Gonzalez, R., and Woods, R., *Digital Image Processing*, Addison Wesley, Reading, 1992.

[66] Goutsias, J., and Schonfeld, D., "Morphological Representation of Discrete and Binary Images," *IEEE Transactions on Signal Processing*, Vol. 39, No. 6, June, 1991.

[67] Goutsias, J., "Morphological Analysis of Discrete Random Shapes," *Journal of Mathematical Imaging and Vision*, Vol. 2, No. 2-3, November, 1992.

[68] Goutsias, J., "Binary Random Fields, Random Set Theory, and the Morphological Analysis of Shape," *Proc. SPIE*, Vol. 2030, pp. 54-64, July, 1993

[69] Grivas, D., and Skolnick, M., "Morphology-Based Image Processing for Pavement Surface Analysis," *Powders and Grains*, eds. Biarez and Gurves, Balkema, Rotterdam, 1989.

[70] Haavisto, P., Gabbouj, M., and Neuvo, Y., "Median Based Idempotent Filters," *Journal of Circuits, Systems and Computers*, Vol. 1, no. 2, pp. 125-148, June 1991.

[71] Hadwiger, H., *Vorslesungen uber Inhalt, Oberflache, and Isoperimetrie*, Springer-Verlag, Berlin, 1957.

[72] Haralick, R. M., Dougherty, E. R., and Katz, P. L., "Model-Based Morphology: The Opening Spectrum" in *Advances in Image Analysis*, ed. Y. Mahdavieh and R. C. Gonzalez, SPIE Press, 1992.

[73] Haralick, R., and Shapiro, L., *Machine Vision*, Addison-Wesley, Boston, 1991.

[74] Haralick, R., Sternberg, S., and Zhuang, X., "Image Analysis Using Mathematical Morphology," *IEEE Transactions on Pattern Analysis and Machine Intelligence*, Vol. 9, No. 4, July, 1987.

[75] Haralick, R., Zhuang, X., Lin, C., and Lee, J., "The Digital Morphological Sampling Theorem," *IEEE Transactions on Acoustics, Speech, and Signal Processing*, Vol. 37, No. 12, December, 1989.

[76] Heijmans, H., "Theoretical Aspects of Gray-Level Morphology," *IEEE Transactions on Pattern Analysis and Machine Intelligence*, Vol. 13, 1991.

[77] Heijmans, H., "Morphological Filtering and Iteration," *Proc. SPIE*, Vol. 1360, pp. 166-175, October, 1990.

[78] Heijmans, H., and Ronse, C., "The Algebraic Basis of Mathematical Morphology, I. Dilations and Erosions," *Computer Vision, Graphics, and Image Processing*, Vol. 50, No. 3, June, 1990.

[79] Heijmans, H., and Ronse, C., "The Algebraic Basis of Mathematical Morphology, II. Openings and Closings," *Computer Vision, Graphics, and Image Processing - Image Understanding*, Vol. 54, 1991.

[80] Heijmans, H., and Vincent, L., "Graph Morphology in Image Analysis," in *Mathematical Morphology in Image Processing*, ed. E. R. Dougherty, Marcel Dekker, New York, 1993.

[81] Heijmans, H., "On the Construction of Morphological Operators which are Selfdual and Activity-Extensive," *Proc. EURASIP Conference on Mathematical Morphology and its Application to Signal Processing*, Barcelona, May, 1993.

[82] Heijmans, H., and Toet, A., "Morphological Sampling," *Journal of Computer Vision, Graphics, and Computer Vision: Image Understanding*, Vol. 54, No. 3, November, 1991.

[83] Heijmans, H., and Serra, J., "Convergence, Continuity, and Iteration on Mathematical Morphology," *Journal of Visual Communication and Image Representation*, Vol. 3, No. 1, March, 1992.

[84] Heinonen, P., and Neuvo, Y., "FIR-Median Hybrid Filters," *IEEE Trans. Acoust., Speech, Signal Processing*, Vol. 35, pp. 832-838, June. 1987.

[85] Hsueh, Yuang-Cheh, "Mathematical Morphology on 1-images," *Journal of Signal Processing*, Vol. 26, No. 2, February, 1992.

[86] Huang, T. S., Yang, G. J., and Tang, G. Y., "A Fast Two-Dimensional Median Filtering Algorithm," *IEEE Trans. Acoust. Speech, Signal Processing*, Vol. 27, pp. 13-18, Feb. 1979.

[87] Huber, P. J., *Robust Statistics*, Wiley, New York, 1981.

[88] Jeulin, D., "Sequential Random Functions," in *Geostatistics*, Vol. 1, ed. M. Armstrong, 1989.

[89] Jeulin, D., "Morphological Modeling of Images by Sequential Random Functions," *Signal Processing*, Vol. 16, No. 4, April, 1989.

[90] Jones, R., and Svalbe, I., "The Design of Morphological Filters Using Multiple Structuring Elements. II. Open(close) and Close(open)," *Pattern Recognition Letters*, Vol. 13, No. 3, March, 1992.

[91] Jones, R., and Svalbe, I., "Basis Decomposition of Morphological Operations," *Proc. 11th IAPR International Conference on Pattern Recognition*, The Hague, August, 1992.

[92] Juhola, M., Katajainen, J., and Raita, T., "Comparison of Algorithms for Standard Median Filtering," *IEEE Trans. Signal Processing*, Vol. 39, pp. 204-208, Jan. 1991.

[93] Justusson, B. I., "Median filtering: Statistical properties," in *Topics in Applied Physics, Two-Dimensional Digital Signal Processing II*, T. S. Huang, Ed. Berlin, Germany: Springer-Verlag, 1981, Vol. 43, pp. 161-196.

[94] Kassam, S. A., and Poor, H. V., "Robust Techniques for Signal Processing: A survey," *Proc. IEEE*, Vol. 73, Mar. 1985.

[95] Klein, J., and Serra, J., "The Texture Analyser," *Journal of Microscopy*, Vol. 95, April, 1973.

[96] Ko, S. J., and Lee, Y. H., "Center Weighted Median Filters and Their Applications to Image Enchangement," *IEEE Trans. Circuits Syst.*, Vol. 38, pp. 984-993, Sept. 1991.

[97] Kong, X., and Goutsias, J., "Comparison of Pyramidal Image Decomposition Techniques for Image Representation and Compression," *Proc. SPIE*, Vol. 2030, pp. 252-265, July, 1993.

[98] Koskinen, L., Astola, J., and Neuvo, Y., "Soft Morphological Filters," *Proc. SPIE*, Vol. 1568, pp. 262-270, July, 1991.

[99] Koskinen, L., and Astola, J., "Morphological Filtering of Noisy Images," *Proc. SPIE*, Vol. 1360, pp. 155-165, October, 1990.

[100] Koskinen, L., Astola, J., and Neuvo, Y., "Statistical Properties of Discrete Morphological Filters," *Proc. IEEE ISCAS-90*.

[101] Koskinen, L., Astola, J., and Neuvo, Y., "Analysis of Noise Attenuation in Morphological Image Processing," *Proc. SPIE*, Vol. 1451, pp. 102-113, February, 1991.

[102] Koskinen, L., and Astola, J., "Statistical Properties of Soft Morphological Filters," *Proc. SPIE*, Vol. 1658, pp. 25-36, February, 1992.

[103] Koskinen, L., and Astola, J., "Asymptotic Behaviour of Morphological Filters," *Journal of Mathematical Imaging and Vision 2*, 117-135, 1992.

[104] Kraus, E., "Domain-variant Gray-scale Morphology," *Proc. SPIE*, Vol. 1451, pp. 171-178, February, 1991.

[105] Kuosmanen, P., and Astola, J., "Selection probabilities", *Image Algebra and Morphological Image Processing IV*, Proc. SPIE, Vol. 2030, pp. 203-217, July 1993.

[106] Lewis, P. M., and Coates, C. L., *Threshold Logic*, John Wiley & Son, New York, 1967.

[107] Lin, J. H., and Coyle, E. J., "Minimum Mean Absolute Error Estimation over the Class of Generalized Stack Filters", *IEEE Trans. Acoust., Speech, Signal Processing*, Vol. 38, pp. 663-678, Apr. 1990.

[108] Lin, J. H., Sellke, T. M., and Coyle, E. J., "Adaptive Stack Filtering under the Mean Absolute Error Criterion", *IEEE Trans. Acoust., Speech, Signal Processing*, Vol. 38, pp. 938-954, June 1990.

[109] Loce, R., and Dougherty, E. R., "Mean-Absolute-Error Theorem for Computational Morphology," *Proc. SPIE*, Vol. 2030, pp. 88-96, July, 1993.

[110] Loce, R. P., and Dougherty, E. R., "Optimal Morphological Restoration: The Morphological Filter Mean-Absolute-Error Theorem," *Journal of Visual Communication and Image Representation*, Vol. 3, No. 4, December, 1992.

[111] Loce, R. P., and Dougherty, E. R., "Facilitation of Optimal Binary Morphological Filter Design Via Structuring-Element Libraries and Observation Constraints," *Optical Engineering*, Vol. 31, No. 5, May, 1992.

[112] Longbotham, H. G., and Bovik, A. C., "Theory of order statistics filters and their relationships to linear FIR filters," *IEEE Trans. Acoust., Speech, Signal Processing*, Vol. 37, pp. 275-287, Feb. 1989.

[113] Longbotham, H. G., and Eberly, D., "The WMMR Filters: A Class of Robust Edge Enhancers," *IEEE Trans. Signal Proc.*, April 1993.

[114] Maragos, P., "Max-Min Difference Equations and Recursive Morphological Systems," *Proc. EURASIP Workshop on Mathematical Morphology and its Applications to Signal Processing*, Barcelona, May, 1993.

[115] Maragos, P., "A Representation Theory for Morphological Image and Signal Processing," *IEEE Transactions on Pattern Analysis and Machine Intelligence*, Vol. 11, No. 6, June, 1989.

[116] Maragos, P., "Morphology-based Symbolic Image Modeling, Multiscale Nonlinear Smoothing, and Pattern Spectrum," *Proc. IEEE Conference on Computer Vision and Pattern Recognition*, June, 1988.

[117] Maragos, P., "Pattern Spectrum and Multi-Scale Shape Representation," *IEEE Transactions on Pattern Analysis and Machine Intelligence*, Vol. 11, July, 1989.

[118] Maragos, P. and Schafer, R., "Morphological Systems for Multidimensional Signal Processing," *Proc. IEEE*, Vol. 78, No. 4, April, 1990.

[119] Maragos, P., and Schafer, R., "A Unification of Linear, Median, Order-Statistics and Morphological Filters under Mathematical Morphology," *Proc. ICASSP*, March, 1985.

[120] Maragos, P., and Schafer, R. W., "Morphological filters-part I: Their Set-Theoretic Analysis and Relations to Linear Shift-Invariant Filters," *IEEE Trans. Acoust., Speech, Signal Processing*, Vol. ASSP-35, pp. 1153-1169, Aug. 1987.

[121] Maragos, P., and Schafer, R. W., "Morphological filters-part II: Their Relations to Median, Order-Statistics, and Stack Filters," *IEEE Trans. Acoust., Speech, Signal Processing*, Vol. ASSP-35, pp. 1170-1184, Aug. 1987.

[122] Maragos, P., and Ziff, R., "Threshold Superposition in Morphological Image Analysis Systems," *IEEE Transactions on Pattern Analysis and Machine Intelligence*, Vol. 12, No. 5, May, 1990.

[123] Matheron, G., *Random Sets and Integral Geometry*, John Wiley, New York, 1975.

[124] Matheron, G., "Dilations on Topological Spaces," in *Image Analysis and Mathematical Morphology*, Vol. 2, ed. J. Serra, Academic Press, New York, 1988.

[125] Matheron, G., "Filters and Lattices," in *Image Analysis and Mathematical Morphology*, Vol. 2, ed. J. Serra, Academic Press, New York, 1988.

[126] Matheron, G., and Serra, J., "Strong Filters and Connectivity," in *Image Analysis and Mathematical Morphology*, Vol. 2, ed. J. Serra, Academic Press, New York, 1988.

[127] Matheron, G., "On the Negligibility of the Skeleton and the Absolute Continuity of Erosions," in *Image Analysis and Mathematical Morphology*, Vol. 2, ed. J. Serra, Academic Press, New York, 1988.

[128] Matheron, G., "Random Sets Theory and its Application to Stereology," *Journal of Microscopy*, Vol. 95, February, 1972.

[129] Mathew, A. V., Dougherty, E. R., and Swarnakar, V., "Efficient Derivation of the Optimal Mean-Square Binary Morphological Filter From the Conditional Expectation Via a Switching Algorithm For the Discrete Power-Set Lattice," *Journal of Circuits, Systems, and Signal Processing*, Vol. 12, No. 3, June, 1993.

[130] Mattioli, J., and Schmitt, M., "Inverse Problems for Granulometries by Erosion," *Journal of Mathematical Imaging and Vision*, Vol. 2, No. 2-3, November, 1992.

[131] Mattioli, J., "Differential Relations of Morphological Operators," *Proc. EURASIP Conference on Mathematical Morphology and its Applications to Signal Processing*, Barcelona, May, 1993.

[132] Mattioli, J., and Schmitt, M., "Efficient Algorithm for Computing the Antigranulometry," *Proc. SPIE*, Vol. 2030, pp. 167-178, July, 1993.

[133] Meyer, F., and Beucher, S., "Morphological Segmentation," *Visual Communication and Image Representation*, Vol. 1, No. 1, September, 1990.

[134] Meyer, F., and Serra, J., "Contrasts and Activity Lattice," *Signal Processing*, Vol. 16, No. 4, April, 1989.

[135] Minkowski, H., "Volumen and Oberflache," *Math. Ann.*, Vol. 57, 1903.

[136] Muroga, S., "Threshold Logic and its Applications", *Wiley Interscience*, New York, 1971.

[137] Nieminen, A., Heinonen, P., and Neuvo, Y., "A New Class of Detail Preserving Filters for Image Processing," *IEEE Trans. Pattern Anal. Machine Intell.*, Vol. PAMI-9, pp. 74-90, Jan. 1987.

[138] Nodes, T. A., and Gallagher, N. C., "Median Filters: Some Modifications and Their Properties," *IEEE Trans. Acoust., Speech, Signal Processing*, Vol. 30, pp. 739-746, Oct. 1982.

[139] Pasian, F., "Sorting Algorithms for Filters Based on Ordered Statistics: Performance Considerations," *Signal Processing*, Vol. 14, no. 3, pp. 287-293, Apr. 1988.

[140] Pitas, I., and Venetsanopoulos, A. N., "Nonlinear Digital Filters," Kluwer Academic Publishers, 1990.

[141] Pitas, I., and Maglara, A., "Range Image Analysis by Using Morphological Signal Decomposition," *Pattern Recognition*, Vol. 24, No. 2,

[142] Pitas, I., and Venetsanopoulos, A., "Morphological Shape Decomposition," *IEEE Transactions on Pattern Analysis and Machine Intelligence*, Vol. 11, No. 7, July, 1989.

[143] Pitas, I., and Venetsanopoulos, A., "Order Statistics in Digital Image Processing," *Proceedings of the IEEE*, Vol. 80, No. 12, pp. 1893-1921.

[144] Preteux, F., "Advanced Mathematical Morphology: From an Algebraic to a Stochastic Point of View," *Proc. SPIE*, Vol. 1769, pp. 44-58, July, 1992.

[145] Preteux, F., and Schmitt, M., "Boolean Texture Analysis and Synthesis," in *Image Analysis and Mathematical Morphology*, Vol. 2, ed. J. Serra, Academic Press, New York, 1988.

[146] Roerdink, J., and Heijmans, H., "Mathematical Morphology for Structures without Translation Symmetry," *Signal Processing*, Vol. 15, No. 3, October, 1988.

[147] Ronse, C., "Why Mathematical Morphology Needs Complete Lattices," *Signal Processing*, Vol. 21, No. 2, October, 1990.

[148] Rustanius, P., Koskinen, L., and Astola, J., "Theoretical and Experimental Analysis of the Effects of Noise in Morphological Image Processing," *Proc. SPIE*, Vol. 1769, pp. 2-13, July, 1992.

[149] Rystrom, L., Katz, P., Haralick, R., and Eggen, C., "Morphological Algorithm Development Case Study: Detection of Shapes in Low-Contrast Gray-Scale Images with Replacement and Clutter Noise," *Proc. SPIE*, Vol. 1658, pp. 76-93, February, 1992.

[150] Safa, F., and Fouzat, G., "Speckle Removal on Radar Imagery Based on Mathematical Morphology," *Signal Processing*, Vol. 16, No. 4, April, 1989.

[151] Salembier, P., and Jaquenoud, L., "Adaptive Morphological Multiresolution Decomposition," *Proc. SPIE*, Vol. 1568, pp. 26-37, July, 1991.

[152] Salembier, P., "Structuring Element Adaptation for Morphological Filters," *Journal of Visual Communication and Image Representation*, Vol. 3, No. 2, June, 1992.

[153] Sand, F., and Dougherty, E. R., "Heterogeneous Granulometries," *Proc. SPIE*, Vol. 2030, pp. 162-166, July, 1993.

[154] Sand, F., and Dougherty, E. R., "Statistics of the Morphological Pattern-Spectrum Moments For a Random-Grain Model," *Journal of Mathematical Imaging and Vision*, Vol. 1, No. 2, July, 1992.

[155] Sand, F., and Dougherty, E. R., "Asymptotic Normality of the Morphological Pattern-Spectrum Moments and Orthogonal Granulometric Generators," *Journal of Visual Communication and Image Representation*, Vol. 3, No. 2, June, 1992.

[156] Schmitt, M., "On Two Inverse Problems in Mathematical Morphology," in *Mathematical Morphology in Image Processing*, ed E. R. Dougherty, Marcel Dekker, New York, 1993.

[157] Schmitt, M., "Estimation of the Density in a Stationary Boolean Model," *Applied Probability*, Vol. 28, September, 1991.

[158] Schonfeld, D., and Goutsias, J., "Optimal Morphological Pattern Restoration from Noisy Binary Images," *IEEE Transactions on Pattern Analysis and Machine Intelligence*, Vol. 13, No. 1, January, 1991.

[159] Schonfeld, D., and Goutsias, J., "Optimal Morphological Filters for Pattern Restoration," *Proc. SPIE*, Vol. 1199, pp. 158-169, November, 1989.

[160] Schonfeld, D., and Goutsias, J., "On the Morphological Representation of Binary Images in a Noisy Environment," *Visual Communication and Image Representation*, Vol. 2, No. 1, March, 1991.

[161] Schonfeld, D., and Goutsias, J., "Parametric Morphological Filters for Pattern Restoration," *Proc. of the Sixth Multidimensional Signal Processing Workshop*, September, 1989.

[162] Serra, J., "Equicontinuous Functions: A Model for Mathematical Morphology," *Proc. SPIE*, Vol. 1769, pp. 252-263, July, 1992.

[163] Serra, J., *Image Analysis and Mathematical Morphology*, Academic Press, 1983.

[164] Serra, J., "Anamorphoses and Function Lattices (Multivalued Morphology)," in *Mathematical Morphology in Image Processing*, ed. E. R. Dougherty, Marcel Dekker, New York, 1993.

[165] Serra, J., editor, *Image Analysis and Mathematical Morphology*, Vol. 2, Academic Press, New York, 1988.

[166] Serra, J., "Mathematical Morphology for Complete Lattices," in *Image Analysis and Mathematical Morphology*, Vol. 2, ed. J. Serra, Academic Press, New York, 1988.

[167] Serra, J., "Mathematical Morphology for Boolean Lattices," in *Image Analysis and Mathematical Morphology*, Vol. 2, ed. J. Serra, Academic Press, New York, 1988.

[168] Serra, J., "Examples of Structuring Functions and Their Uses," in *Image Analysis and Mathematical Morphology*, Vol. 2, ed. J. Serra, Academic Press, New York, 1988.

[169] Serra, J., "Introduction to Morphological Filters," in *Image Analysis and Mathematical Morphology*, Vol. 2, ed. J. Serra, Academic Press, New York, 1988.

[170] Serra, J., "The Centre and Self-Dual Filtering," in *Image Analysis and Mathematical Morphology*, Vol. 2, ed. J. Serra, Academic Press, New York, 1988.

[171] Serra, J., "Dilation and Filtering for Numerical Functions," in *Image Analysis and Mathematical Morphology*, Vol. 2, ed. J. Serra, Academic Press, New York, 1988.

[172] Serra, J., "Alternating Sequential Filters," in *Image Analysis and Mathematical Morphology*, Vol. 2, ed. J. Serra, Academic Press, New York, 1988.

[173] Serra, J., "Measurements on Numerical Functions," in *Image Analysis and Mathematical Morphology*, Vol. 2, ed. J. Serra, Academic Press, New York, 1988.

[174] Serra, J., "Boolean Random Functions," in *Image Analysis and Mathmatical Morphology*, Vol. 2, ed. J. Serra, Academic Press, New York, 1988.

[175] Serra, J., "Morphological Optics," *Journal of Microscopy*, Vol. 145, January, 1987.

[176] Serra, J., "Introduction to Mathematical Morphology," *Computer Vision, Graphics, and Image Processing*, Vol. 35, No. 3, September, 1986.

[177] Serra, J., "Principles, Criteria and Algorithms in Mathematical Morphology," *Proc. of the NATO Advanced Study Institute on Digital Image Processing and Analysis*, 1980.

[178] Serra, J., "The Boolean Model and Random Sets," *Computer Graphics and Image Processing*, Vol. 12, 1980.

[179] Serra, J., "Links: Definition and Properties," *Proc. SPIE*, Vol. 1360, pp. 202-214, October, 1990.

[180] Serra, J., "Lipschitz Lattices and Numerical Morphology," *Proc. SPIE*, Vol. 1568, pp. 54-65, July, 1991.

[181] Serra, J., and Lay, B., *Algorithms in Mathematical Morphology*, Academic Press, London, 1988.

[182] Serra, J., and Vincent, L., *Lecture Notes on Morphological Filtering*, Ecole Nationale Superieure des Mines de Paris, Fontainebleau, 1989.

[183] Serra, J., and Vincent, L., "An Overview of Morphological Filtering," *Circuits, Systems and Signal Processing*, Vol. 11, No. 1, 1992.

[184] Serra, J., and P. Salembier, "Connected Operators and Pyramids," *Proc. SPIE*, Vol. 2030, pp. 65-77, July, 1993.

[185] Shih, F., and Mitchell, O., "Threshold Decomposition of Gray-Scale Morphology into Binary Morphology," *IEEE Transactions on Pattern Analysis and Machine Intelligence*, Vol. 11, No. 1, January, 1989.

[186] Sideropoulos, N. D., Baras, J. S., and Berenstein, C. A., "Discrete Random Sets: An Inverse Problem, Plus Tools for the Statistical Inference of the Discrete Boolean Model," *Proc. SPIE*, Vol. 1769, pp. 32-43, July, 1992.

[187] Sinha, D., and Dougherty, E. R., "Characterization of Fuzzy Minkowski Algebra," *Proc. SPIE*, Vol. 1769, pp. 59-71, July, 1992.

[188] Sinha, D., and Dougherty, E. R., "Fuzzy Mathematical Morphology," *Journal of Visual Communication and Image Representation*, Vol. 3, No. 3, September, 1992.

[189] Sinha, D., and Dougherty, E. R., "Fuzzification of Set Inclusion: Theory and Applications," *Journal of Fuzzy Sets and Systems*, Vol. 55, No. 1, April, 1993.

[190] Sinha, D., and Giardina, C., "Discrete Black and White Object Recognition via Morphological Functions," *IEEE Transactions on Pattern Analysis and Machine Intelligence*, Vol. 12, No. 3, March, 1990.

[191] Sinha, D., Sinha, P., and Dougherty, E. R., "Algorithm Development for Fuzzy Mathematical Morphology," *Proc. SPIE*, Vol. 2030, pp. 220-230, July, 1993.

[192] Skolnick, M., "Application of Morphological Transformations to the Analysis of Two-dimensional Electrophoretic Gels of Biological Materials," *Computer Vision, Graphics, and Image Processing*, Vol. 35, No. 3, September, 1986.

[193] Soille, P., "Spatial Distributions from Contour Lines: An Efficient Methodology Based on Distance Transformations," *Visual Communication and Image Representation*, Vol. 2, No. 2, June, 1991.

[194] Soille, P., Rivest, J., and Beucher, S., "Morphological Gradients," *Proc. SPIE*, Vol. 1658, pp. 139-150, February, 1992.

[195] Song, J., and Delp, E., "Statistical Analysis of Morphological Operators," *Proc. Twenty Fifth Annual Conference on Information Sciences and Systems*, March, 1991.

[196] Song, J., and Delp, E. J., "A Study of the Generalized Morphological Filter," *Journal of Circuits, Systems, and Signal Processing*, Vol. 11, No. 1, 1992.

[197] Sternberg, S., "Grayscale Morphology," *Computer Vision, Graphics, and Image Processing*, Vol. 35, No. 3, September, 1986.

[198] Sternberg, S., "Cellular Computers and Biomedical Image Processing," *Biomedical Images and Computers*, ed. J. Sklansky and J. Bisconte, Springer-Verlag, Berlin, 1982.

[199] Sternberg, S., "Industrial Morphology," *Proc. SPIE*, Vol. 504, pp. 202-214, August, 1984.

[200] Sternberg, S., "Morphological Cellular Logic Image Processor Architectures," *Proc. SPIE*, Vol. 435, pp. 112-120, August, 1983.

[201] Sternberg, S., "Biomedical Image Processing," *Computer*, ol. 16, No. 1, 1983.

[202] Stevenson, R., and Arce, G., "Morphological Filters: Statistics and Further Syntactic Properties," *IEEE Transactions on Circuits and Systems*, Vol. 34, No. 11, November, 1987.

[203] Svalbe, I., "The Geometry of Basis Sets for Morphologic Closing," *IEEE Transactions on Pattern Analysis and Machine Intelligence*, Vol. 13, No. 12, December, 1991.

[204] Svalbe, I., and Jones, R., "The Design of Morphological Filters Using Multiple Structuring Elements. I. Openings and Closings," *Pattern Recognition Letters*, Vol. 13, No. 2, February, 1992.

[205] Tyan, S. G., "Median Filtering: Deterministic Properties," in *Topics in Applied Physics, Two-Dimensional Digital Signal Processing II*, T. S. Huang, Ed. Berlin, Germany: Springer-Verlag, 1981, Vol. 43, pp. 197-217.

[206] Vincent, L., "Grayscale Area Openings and Closings, Their Efficient Implementation and Applications," *Proc. EURASIP Workshop on Mathematical Morphology and its Applications to Signal Processing*, Barcelona, May, 1993.

[207] Vincent, L., "Morphological Transformations of Binary Images with Arbitrary Structuring Elements," *Signal Processing*, Vol. 22, No. 1, January, 1991.

[208] Vincent, L., "Graphs and Mathematical Morphology," *Signal Processing*, Vol. 16, No. 4, April, 1989.

[209] Vincent, L., "Algorithmes Morphologiques a Base de Files d'Attente et de Lacets. Extension aux Graphes," Ph. D. Theses, *Ecole des Mines de Paris*, May, 1990.

[210] Weisman, A., Dougherty, E. R., Mises, H., and Miller, D., "Nonlinear Digital Filtering of Scanning-Probe-Microscopy Images by Morphological Pseudoconvolutions," *Journal of Applied Physics*, Vol. 71, No. 4, February, 1992.

[211] Wendt, P. D., Coyle, E. J., and Gallagher, N. C., Jr., "Stack filters," *IEEE Trans. Acoust., Speech, Signal Processing*, Vol. ASSP-34, pp. 898-911, Aug. 1986.

[212] Wilson, S., "Theory of Matrix Morphology," *IEEE Transactions on Pattern Analysis and Machine Intelligence*, Vol. 14, No. 6, June, 1992.

[213] Wilson, S., "Floating Stack Arrays: A Unified Representation of Linear and Morphological Filters," *Proc. SPIE*, Vol. 1769, pp. 332-343, July, 1992.

[214] Wilson, S., "Vector Morphology and Iconic Neural Networks," *IEEE Transactions on Systems, Man, and Cybernetics*, Vol. 19, No. 6, November, 1989.

[215] Wilson, S., "Applications of Matrix Morphology," *Proc. SPIE*, Vol. 1350, pp. 44-55, July, 1990.

[216] Wilson, S., "Morphological Networks," *Proc. SPIE*, Vol. 1199, pp. 483-493, November, 1989.

[217] Wilson, S., "Order-statistic Filters on Matrices of Images," *Proc. SPIE*, Vol. 1451, pp. 242-253, February, 1991.

[218] Wilson, S., "Training Structuring Elements in Morphological Networks," in *Mathematical Morphology in Image Processing*," ed. E. R. Dougherty, Marcel Dekker, New York, 1993.

[219] Wilson, S., "Training Object Classes Using Mathematical Morphology," *Proc. SPIE*, Vol. 1658, pp. 267-275, February, 1992.

[220] Yang, R., Yin, L., Gabbouj, M., Astola, J., and Neuvo, Y,, "Optimal Weighted Median Filters under Structural Constraints", *IEEE Trans. Signal Processing*, submitted.

[221] Yin, L., Astola, J., and Neuvo, Y,, "Adaptive Stack Filtering with Applications to Image Processing", *IEEE Trans. Signal Processing*, Vol. 41, No. 1, pp. 162-184, January 1983.

[222] Yli-Harja, O., Astola, J., and Neuvo, Y., "Analysis of the Properties of Median and Weighted Median Filters Using Threshold Logic and Stack Filter Representation, " *IEEE Trans. Signal Processing*, Vol. SP-39, pp. 395-410, Feb. 1991.

[223] Yuille, A., and Vincent, L., "Statistical Morphology and Bayesian Reconstruction," *Journal of Mathematical Imaging and Vision*, Vol. 1, No. 3, 1992.

[224] Yuille, A., Vincent, L., and Geiger, D., "Statistical Morphology," *Proc. SPIE*, Vol. 1568, pp. 271-282, July, 1991.

[225] Zhao, D., and Daut, D., "Morphological Hit-or-Miss Transformation for Shape Recognition," *Visual Communication and Image Representation*, Vol. 2, No. 3, September, 1991.

Index

Edward R. Dougherty is a professor at the Center for Imaging Science of the Rochester Institute of Technology and also serves as an industrial consultant. He holds an M.S. in computer science from Stevens Institute of Technology and a Ph.D. in mathematics from Rutgers University. He has written numerous papers in the areas of mathematical morphology, image algebra, and nonlinear image processing. Including the present one, he has authored/coauthored seven books, the others being *Matrix Structured Image Processing, Image Processing — Continuous to Discrete, Morphological Methods in Image and Signal Processing, Mathematical Methods for Artificial Intelligence and Autonomous Systems, Probability and Statistics for the Engineering, Computing, and Physical Sciences*, and *An Introduction to Morphological Image Processing*. He has also edited three books, *Mathematical Morphology in Image Processing, Digital Image Processing Methods*, and *Mathematical Nonlinear Image Processing*, the latter with J. Astola. He regularly teaches short courses in morphological image processing, including those for SPIE, serves as a chair for two SPIE conferences, Image Algebra and Morphological Image Processing, and Nonlinear Image Processing, and currently serves as chair of the SPIE Working Group on Electronic Imaging.

Jaakko Astola was born in Helsinki, Finland, on May 6, 1949. He received the B.Sc., M.Sc., Licentiate, and Ph.D. degrees in mathematics from Turku University, Finland, in 1972, 1973, 1975, and 1978, respectively. From 1976 to 1977 he was a research assistant at the Research Institute for Mathematical Sciences of Kyoto University, Kyoto, Japan. Between 1979 and 1987 he was with the Department of Information Technology, Lappeenranta University of Technology, Lappeenranta, Finland, holding various teaching positions in mathematics, applied mathematics and computer science. From 1988 to 1993 he was an Associate Professor in applied mathematics at Tampere University, Tampere, Finland. Currently he is a Professor of digital signal processing at Tampere University of Technology. His research interests include image and signal processing, coding theory, and statistics.